User Guide for MODPATH Version 6—
A Particle-Tracking Model for MODFLOW

By David W. Pollock

Chapter 41 of
Section A, Groundwater
Book 6, Modeling Techniques

Techniques and Methods 6–A41

U.S. Department of the Interior
U.S. Geological Survey

U.S. Department of the Interior
KEN SALAZAR, Secretary

U.S. Geological Survey
Marcia K. McNutt, Director

U.S. Geological Survey, Reston, Virginia: 2012

For more information on the USGS—the Federal source for science about the Earth, its natural and living resources, natural hazards, and the environment, visit http://www.usgs.gov or call 1–888–ASK–USGS.

For an overview of USGS information products, including maps, imagery, and publications,
visit http://www.usgs.gov/pubprod

To order this and other USGS information products, visit http://store.usgs.gov

Suggested citation:
Pollock, D.W., 2012, User Guide for MODPATH Version 6—A Particle-Tracking Model for MODFLOW: U.S. Geological Survey Techniques and Methods 6–A41, 58 p.

Preface

This report describes MODPATH, a particle-tracking model designed to work with output from the U.S. Geological Survey groundwater flow model commonly referred to as MODFLOW. The model program can be obtained at http://water.usgs.gov/software/ground_water.html/. The performance of the program has been tested for a variety of test problems; however, future applications might reveal errors that were not detected in the test simulations. Users should send notification of any errors found in the report or the model program to:

Office of Groundwater
U.S. Geological Survey
Mail Stop 411
12201 Sunrise Valley Drive
Reston, VA 20192
(703) 648-5001

Updates might be made both to the report and the model program. Users can check for updates at the above Internet address.

Contents

Figures

Conversion Factors

Inch/Pound to SI

Multiply	By	To obtain
foot (ft)	0.3048	meter (m)
foot per day (ft/d)	0.3048	meter per day (m/d)
cubic foot per day (ft^3/d)	0.02832	cubic meter per day (m^3/d)

Abbreviations

LGR	Local Grid Refinement (feature)
LPF	Layer Property Flow (package)
MPBAS	MODPATH Basic Data (file)
PDF	Portable Document Format
USGS	U.S. Geological Survey

User Guide for MODPATH Version 6—A Particle-Tracking Model for MODFLOW

By David W. Pollock

Abstract

MODPATH is a particle-tracking post-processing model that computes three-dimensional flow paths using output from groundwater flow simulations based on MODFLOW, the U.S. Geological Survey (USGS) finite-difference groundwater flow model. This report documents MODPATH version 6. Previous versions were documented in USGS Open-File Reports 89-381 and 94-464.

The program uses a semianalytical particle-tracking scheme that allows an analytical expression of a particle's flow path to be obtained within each finite-difference grid cell. A particle's path is computed by tracking the particle from one cell to the next until it reaches a boundary, an internal sink/source, or satisfies another termination criterion.

Data input to MODPATH consists of a combination of MODFLOW input data files, MODFLOW head and flow output files, and other input files specific to MODPATH. Output from MODPATH consists of several output files, including a number of particle coordinate output files intended to serve as input data for other programs that process, analyze, and display the results in various ways.

MODPATH is written in FORTRAN and can be compiled by any FORTRAN compiler that fully supports FORTRAN-2003 or by most commercially available FORTRAN-95 compilers that support the major FORTRAN-2003 language extensions.

Introduction

MODPATH is a particle-tracking post-processing program designed to work with the U.S. Geological Survey (USGS) finite-difference groundwater flow model, commonly known as MODFLOW (Harbaugh, 2005). Output from steady-state or transient MODFLOW simulations is used in MODPATH to compute paths for imaginary "particles" of water moving through the simulated groundwater system. In addition to computing particle paths, MODPATH computes the time of travel for particles moving through the system. By carefully defining the starting locations of particles, it is possible to perform a wide range of flow-system analyses, such as delineating recharge areas and drawing flow nets.

This report describes MODPATH version 6, which is the third major MODPATH release since its publication in original form (Pollock, 1989). The original version of MODPATH was restricted to steady-state MODFLOW simulations. The second major release (Pollock, 1994) added support for transient flow simulations, as well as many other new features. To an experienced MODPATH user, this new release might appear to be more evolutionary than revolutionary because relatively few of its features are totally new. Many preexisting features have been improved, however, and the computer code has been modernized in a manner that greatly improves its structure, organization, and ability (unlike previous versions) to support new features in the future.

This version of MODPATH has been rewritten using the same coding paradigm as MODFLOW-2005 to provide the infrastructure to support multiple grids and local grid refinement (Mehl and Hill, 2005). Although MODPATH version 6 still is restricted to a single finite-difference grid, the code structure and data file formats are in place to extend support to multiple grids in the future.

MODPATH version 6 improves performance for transient flow simulations. In particular, the composite budget file required by previous versions for transient particle tracking is no longer required. Data input has also been simplified; the built-in text-based preprocessor present in previous versions has been replaced with a simpler data file that is easier for data pre-processing applications to support. Lastly, model output has been redesigned to provide more information for post-processing applications.

This report is organized into five major sections:

- Method
- MODPATH Overview
- Input Instructions
- Examples
- Appendix—MODPATH Output Examiner

Method focuses primarily on describing the semianalytical tracking algorithm that is at the core of MODPATH. A thorough understanding of the tracking method is essential to use MODPATH correctly. *MODPATH Overview* discusses several key concepts and provides a general description of the model's input and output. Detailed descriptions of the input and output files are provided in *Input Instructions*. *Examples* presents sample simulations that illustrate data files and demonstrate the steps required to execute a MODPATH simulation. Because particle-tracking output is difficult to evaluate without the aid of visual display tools, a simple Microsoft-Windows-based application that displays and examines particle-tracking output is described in *Appendix—MODPATH Output Examiner*.

Method

Semianalytical Tracking Algorithm

The particle-tracking algorithm used by MODPATH can be implemented for either steady-state or transient flow fields. For simplicity, the algorithm is first described for steady-state flow and then extended to transient flow systems. The partial differential equation describing conservation of mass in a steady-state, three-dimensional groundwater flow system can be expressed as

$$\frac{\partial}{\partial x}(nv_x) + \frac{\partial}{\partial y}(nv_y) + \frac{\partial}{\partial z}(nv_z) = W ,$$ (1)

where v_x, v_y, and v_z are the principal components of the average linear groundwater velocity vector, n is porosity, and W is the volume rate of water created or consumed by internal sources and sinks per unit volume of aquifer. Equation 1 expresses conservation of mass for an infinitesimally small volume of aquifer. The finite-difference approximation of equation 1 can be thought of as a mass balance equation for a finite-sized cell of aquifer volume that accounts for water flowing into and out of the cell, and for water generated or consumed within the cell. Figure 1 shows a finite-sized cell of aquifer volume and the components of inflow and outflow across its six faces.

In the discussion that follows, the six cell faces are referred to as x_1, x_2, y_1, y_2, z_1, and z_2. Face x_1 is the face perpendicular to the x-direction at $x = x_1$. Similar definitions hold for the other five faces. The average linear velocity component across each face in cell (i, j, k) is obtained by dividing the volumetric flow rate across the face by the cross sectional area of the face and the porosity of the material contained in the cell,

$$v_{x_1} = \frac{Q_{x_1}}{(n\Delta y\Delta z)}, \quad v_{x_2} = \frac{Q_{x_2}}{(n\Delta y\Delta z)}$$ (2a, 2b)

$$v_{y_1} = \frac{Q_{y_1}}{(n\Delta x\Delta z)}, \quad v_{y_2} = \frac{Q_{y_2}}{(n\Delta x\Delta z)}$$ (2c, 2d)

$$v_{z_1} = \frac{Q_{z_1}}{(n\Delta x\Delta y)}, \quad v_{z_2} = \frac{Q_{z_2}}{(n\Delta x\Delta y)} ,$$ (2e, 2f)

where the Q terms are the volumetric flow rates across the six cell faces and Δx, Δy, and Δz are the dimensions of the cell in the respective coordinate directions. If flow to internal sources or sinks within the cell is specified as Q_s, the following mass balance equation can be applied to the cell:

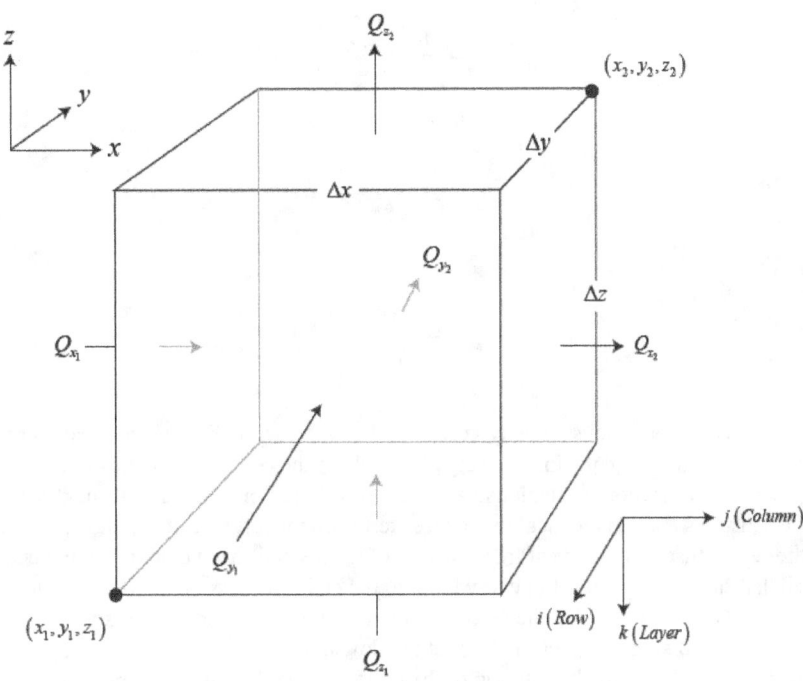

Figure 1. Finite-difference cell showing the definitions of the *x-y-z* coordinates and the *i-j-k* grid indices.

$$\frac{\left(nv_{x_2} - nv_{x_1}\right)}{\Delta x} + \frac{\left(nv_{y_2} - nv_{y_1}\right)}{\Delta y} + \frac{\left(nv_{z_2} - nv_{z_1}\right)}{\Delta z} = \frac{Q_s}{\Delta x \Delta y \Delta z} \ . \tag{3}$$

The left side of equation 3 represents the net volumetric rate of outflow per unit volume of the cell, and the right side represents the net volumetric rate of production per unit volume from internal sources and sinks. Substitution of Darcy's Law for each of the flow terms in equation 3 results in a set of algebraic equations expressed in terms of heads at nodes located at the cell centers. The solution of that set of algebraic equations yields the values of head at the node points. Once the head solution has been obtained, the intercell flow rates can be computed from Darcy's Law using the values of head at the node points. The USGS modular three-dimensional finite-difference groundwater flow model, commonly known as MODFLOW, solves for head and calculates intercell flow rates (Harbaugh, 2005).

In order to compute pathlines, a method must be established to compute values of the principal components of the velocity vector at every point in the flow field based on the intercell flow rates from the finite difference model. The algorithm described in this report uses simple linear interpolation to compute the principal velocity components at points within a cell. Simple linear interpolation allows the velocity components to be expressed in the form

$$v_x = A_x \left(x - x_1\right) + v_{x_1} \tag{4a}$$

$$v_y = A_y \left(y - y_1\right) + v_{y_1} \tag{4b}$$

$$v_z = A_z \left(z - z_1\right) + v_{z_1} \tag{4c}$$

where A_x, A_y, and A_z are constants that correspond to the components of the velocity gradient within the cell,

$$A_x = \frac{\left(v_{x_2} - v_{x_1}\right)}{\Delta x} \quad\quad (5a)$$

$$A_y = \frac{\left(v_{y_2} - v_{y_1}\right)}{\Delta y} \qu\quad (5b)$$

$$A_z = \frac{\left(v_{z_2} - v_{z_1}\right)}{\Delta z}. \quad\quad (5c)$$

Linear interpolation produces a continuous velocity vector field within each individual cell that identically satisfies the differential conservation of mass equation (equation 1) everywhere within the cell. That point can be illustrated by noting that when the linear velocity component functions (equations 4a-4c) are substituted into equation 1, the three derivatives on the left side of equation 1 become constants that are identical to the three terms on the left side of equation 3 if porosity is considered to be constant throughout the cell. Consequently, linear interpolation of the six cell-face velocity components results in a velocity vector field within the cell that satisfies equation 1 at every point in the cell if it is assumed that internal sources and sinks are uniformly distributed within the cell. The fact that the velocity vector field within each cell satisfies the differential mass balance equation assures that pathlines will distribute water throughout the flow field in a way that is consistent with the overall movement of water in the system as represented by the solution of the finite-difference flow equations.

Consider the movement of a particle, p, through a three-dimensional finite-difference cell. The rate of change in the particle's x-component of velocity as it moves through the cell is given by

$$\left(\frac{dv_x}{dt}\right)_p = \left(\frac{dv_x}{dx}\right)\left(\frac{dx}{dt}\right)_p. \quad\quad (6)$$

To simplify notation, the subscript, p, is used to indicate that a term is evaluated at the location of the particle denoted by the spatial coordinates, $\left(x_p, y_p, z_p\right)$. For example, the term $\left(\frac{dv_x}{dt}\right)_p$ is the time rate of change in the x-component of velocity evaluated at the location of the particle. In equation 6, the term $\left(\frac{dv_x}{dx}\right)$ is the rate of change in the x-component of velocity with respect to the x-coordinate. The term $\left(\frac{dx}{dt}\right)_p$ is the time rate of change of the x-location of the particle and, by definition, is the velocity component of the particle in the x-direction at its current location (denoted by the subscript p),

$$v_{x_p} = \left(\frac{dx}{dt}\right)_p. \quad\quad (7)$$

Differentiating equation 4a with respect to x yields the additional relation

$$\frac{dv_x}{dx} = A_x. \qu\quad (8)$$

Substituting equations 7 and 8 into equation 6 gives

$$\left(\frac{dv_x}{dt}\right)_p = A_x v_{x_p}. \qu\quad (9a)$$

Analogous equations are obtained for the y and z directions:

$$\left(\frac{dv_y}{dt}\right)_p = A_y v_{y_p} \tag{9b}$$

$$\left(\frac{dv_z}{dt}\right)_p = A_z v_{z_p} \;. \tag{9c}$$

Equations 9a through 9c can be rearranged to the form

$$\frac{d\left(v_{x_p}\right)}{v_{x_p}} = A_x \Delta t \;. \tag{10}$$

Equation 10 can be integrated and evaluated from time t_1 to t_2 (where $t_2 > t_1$) to yield

$$\ln\left[\frac{\left(v_{x_p}\right)_{t_2}}{\left(v_{x_p}\right)_{t_1}}\right] = A_x \Delta t \tag{11}$$

where $\Delta t = t_2 - t_1$. By taking the exponential of each side of equation 11, substituting equation 4a for $\left(v_{x_p}\right)_{t_2}$, and rearranging, we obtain

$$\left(x_p\right)_{t_2} = x_1 + \frac{1}{A_x}\left[\left(v_{x_p}\right)_{t_1} e^{\left(A_x \Delta t - v_{x_1}\right)}\right]. \tag{12a}$$

Analogous equations can be developed for the y and z directions:

$$\left(y_p\right)_{t_2} = y_1 + \frac{1}{A_y}\left[\left(v_{y_p}\right)_{t_1} e^{\left(A_y \Delta t - v_{y_1}\right)}\right] \tag{12b}$$

$$\left(z_p\right)_{t_2} = z_1 + \frac{1}{A_z}\left[\left(v_{z_p}\right)_{t_1} e^{\left(A_z \Delta t - v_{z_1}\right)}\right]. \tag{12c}$$

The velocity components of the particle at time t_1 are known functions of the particle's coordinates; therefore, the coordinates of the particle at any future time, t_2, can be computed directly from equations 12a through 12c.

For steady-state flow, the direct integration method described above can be imbedded in a simple algorithm that allows a particle's exit point from a cell to be determined directly given any known starting location within the cell. To illustrate the method, consider the two-dimensional example shown in figure 2. Cell (i, j) is in the x-y plane and contains a particle, p, located at $\left(x_p, y_p\right)$ at time t_p. For this example, it is assumed that v_{x_1} and v_{x_2} are greater than zero; that is, water flows into the cell across face x_1 and out of the cell across face x_2. Similarly, it is assumed that v_{y_1} and v_{y_2} are also greater than zero, so that water flows into the cell across face y_1 and out of the cell across face y_2.

The first step is to determine the face across which the particle leaves the cell. For this example, that is accomplished by noting that the velocity components at the four faces require that the particle leave the cell through either face x_2 or face y_2. Considering the x-direction first, the velocity v_{x_p} can be calculated at point $\left(x_p, y_p\right)$ from equation 4a. Because v_x is equal to v_{x_2} at face x_2, equation 11 can be used to determine the time required for the particle to reach face x_2:

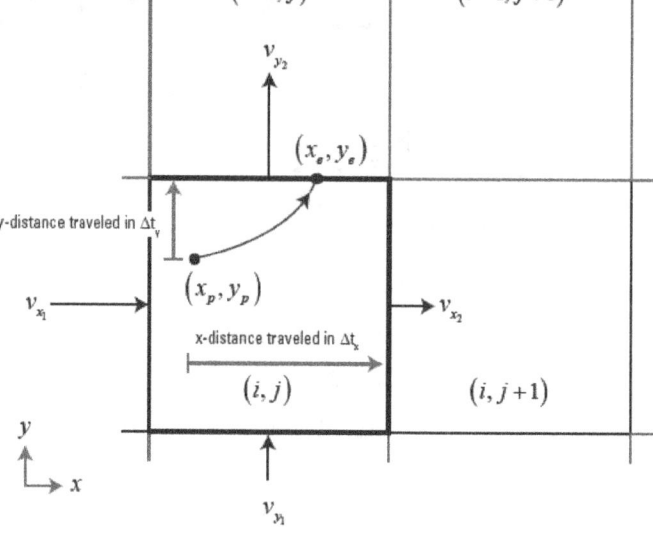

Figure 2. Computation of the exit point and travel time for the case of two-dimensional flow in the x-y plane.

$$\Delta t_x = \frac{1}{A_x} \ln\left(\frac{v_{x_2}}{v_{x_p}}\right)$$

(13a)

An analogous calculation can be made to determine the time required for the particle to reach face y_2,

$$\Delta t_y = \frac{1}{A_y} \ln\left(\frac{v_{y_2}}{v_{y_p}}\right),$$

(13b)

where v_{x_p} and v_{y_p} are the velocity components of the particle at location (x_p, y_p). If Δt_x is less than Δt_y, the particle will leave the cell across face x_2 and enter cell (i, j+1). If Δt_x is greater than Δt_y, the particle will leave the cell across face y_2 and enter cell (i – 1, j). A third possibility is that the two computed exit times are equal, in which case the particle would leave through the corner of cell (i, j) and enter cell (i – 1, j + 1). The particle trajectory shown in figure 2 corresponds to a situation in which Δt_y is less than Δt_x. The length of time required for the particle to travel from point (x_p, y_p) to a boundary face of the cell is taken to be the smaller of Δt_x and Δt_y, and is denoted as Δt_e. The value of Δt_e is then used in equations 12a through 12c to determine the exit coordinates of the particle, (x_e, y_e):

$$x_e = x_1 + \frac{1}{A_x}\left[\left(v_{x_p}\right)_{t_p} e^{\left(A_x \Delta t - v_{x_1}\right)}\right]$$

(14a)

$$y_e = y_1 + \frac{1}{A_y}\left[\left(v_{y_p}\right)_{t_p} e^{\left(A_y \Delta t - v_{y_1}\right)}\right].$$

(14b)

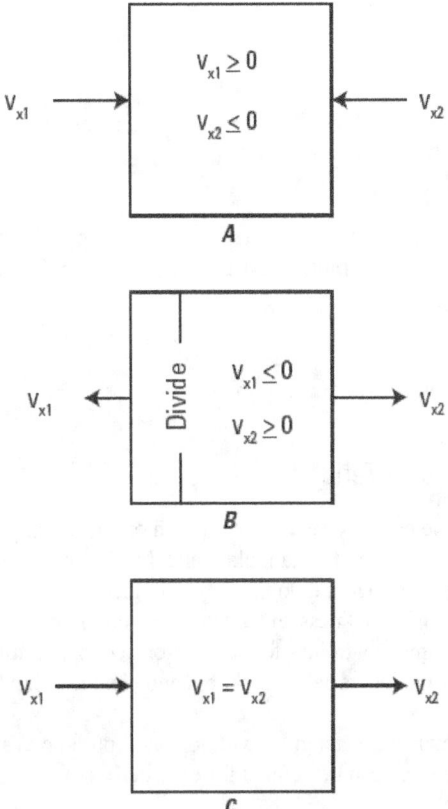

Figure 3. Additional combinations of velocities between pairs of cell faces.

The time at which the particle leaves the cell is given by $t_e = t_p + \Delta t_e$. This sequence of calculations is repeated, cell by cell, until the particle reaches a discharge point. The approach can be generalized to three dimensions in a straightforward manner by solving an analogous set of calculations for the z-direction.

It often is desirable to calculate the location of a particle at a specific point in time that does not generally correspond to the time at which the particle passes from one cell to another. For example, one might want to know how far a particle would travel during a specified time interval. In such cases, the coordinates of a particle at an intermediate time can be computed directly from equations 12a through 12c using an appropriate value of Δt within the range of 0 to Δt_e. The method described herein does not require time steps smaller than Δt_e because pathlines are integrated analytically with respect to time. The use of time steps smaller than Δt_e is simply for convenience. The ultimate termination point and time of travel will be the same.

For the purpose of illustration, the preceding example considered a specific case in which all of the velocity components at the cell faces were non-zero and in the positive x and y directions. Of course, those conditions will not always exist; figure 3 illustrates the other possible situations that can occur in any of the three coordinate directions. Figure 3A shows the case in which v_{x_1} and v_{x_2} are in opposite directions and flow is into the cell through both faces x_1 and x_2. For this case, it is apparent that once a particle enters the cell, it cannot leave the cell in the x-direction. When implementing this algorithm, a check is made to determine if this condition exists for a given coordinate direction. If it does, a flag is set to indicate that a particle cannot leave the cell across either of the faces in that direction. When this situation prevails in all three directions, it indicates that a strong sink is present within the cell and no outflow can occur. Figure 3B shows a second alternative in which v_{x_1} and v_{x_2} are in opposite directions and flow is out of the cell through faces x_1 and x_2. This condition implies that a local flow divide exists for the x-direction somewhere within the cell. For this situation, the potential exit face in the x-direction is determined by checking the sign of v_{x_p}. If the sign is negative, the particle has the potential to leave the cell across face x_1; if the sign is positive, the particle has the potential to leave the cell across face x_2. Once the appropriate potential exit face has been determined for a coordinate direction, the transit time of the particle in that direction can be computed as described in the preceding discussion. Lastly, the case in which v_{x_p} is constant and non-zero (figure 3C) must also be considered special because the analytical expression in equation 13a is undefined and cannot be evaluated. In such instances, equation 13a is bypassed and the transit time in the x-direction is computed from the following simple relations:

$$\text{If } v_{x_1} > 0 \text{ then } \Delta t_x = \frac{x_2 - x_p}{v_{x_1}} \quad \text{Or, if } v_{x_1} < 0 \text{ then } \Delta t_x = \frac{x_1 - x_p}{v_{x_1}}. \tag{15a, 15b}$$

Special Cases

The basic algorithm described in the preceding section has been adapted in the computer program MODPATH to deal with several special cases that commonly arise in simulations of real groundwater flow systems:

- Grids with non-rectangular vertical discretization
- Water-table layers
- Quasi three-dimensional representation of confining layers

Non-Rectangular Vertical Discretization

The development presented in the previous section was based on the assumption that the flow domain was discretized into a three-dimensional rectangular grid of horizontal rectangular cells. In practice, however, many three-dimensional finite-difference simulations use a rectangular grid in the horizontal domain and a deformed grid in the vertical direction to allow grid cells to conform to stratigraphic units that vary in thickness and are not perfectly horizontal. The particle-tracking algorithm described above can be used to compute approximate pathlines for such deformed, or "stratigraphic," three-dimensional grids. Figure 4 shows how a simple confined aquifer having variable thickness and elevation can be represented by a vertically deformed finite-difference grid.

Cells are assumed to be horizontal and rectangular with top and bottom elevations equal to the top and bottom elevations of the cell at its node. A local coordinate, z_L, can be defined for each cell as

$$z_L = \frac{z - z_1}{z_2 - z_1} \tag{16}$$

where z_1 and z_2 are the elevations of the bottom and top of the cell, respectively. According to equation 16, the local z-coordinate varies from 0 at the bottom of the cell to 1 at the top of the cell. When a particle is transferred laterally from one cell to another, its local z-coordinate remains the same. Using this approach, if a particle leaves a cell at a position halfway between the top and bottom of the cell, it is assumed to enter the neighboring cell half way between the top and bottom of that cell regardless of how the thickness or absolute elevation of the layer changes from one cell to the next. This procedure is illustrated schematically for the case of lateral flow in a confined aquifer of variable thickness and dip (fig. 4A–B). When all layers are constant in thickness and horizontal, this approach reduces to the algorithm developed above for true rectangular grids (fig. 4C).

The advantage of a stratigraphic three-dimensional grid is that complex hydrogeologic systems can be simulated with fewer layers than would be necessary to adequately represent them using a true rectangular three-dimensional grid. The principal disadvantage is that spatial discretization errors are introduced that are difficult to quantify, especially with respect to pathline computations. Nevertheless, experience has shown that the method described above produces results that are internally consistent with water budgets generated by MODFLOW for models that use deformed vertical grids.

Water-Table Layers

For water-table layers, the saturated thickness of cells changes in relation to the slope of the water table and the bottom elevation of cells within the layer. The top elevation of a cell in a water-table layer is set equal to the head in the cell. Consequently, water-table layers vary in thickness, even for true three-dimensional rectangular grids. The particle-tracking algorithm treats water-table layers in the same way as the variable thickness deformed grid layers described in the preceding section.

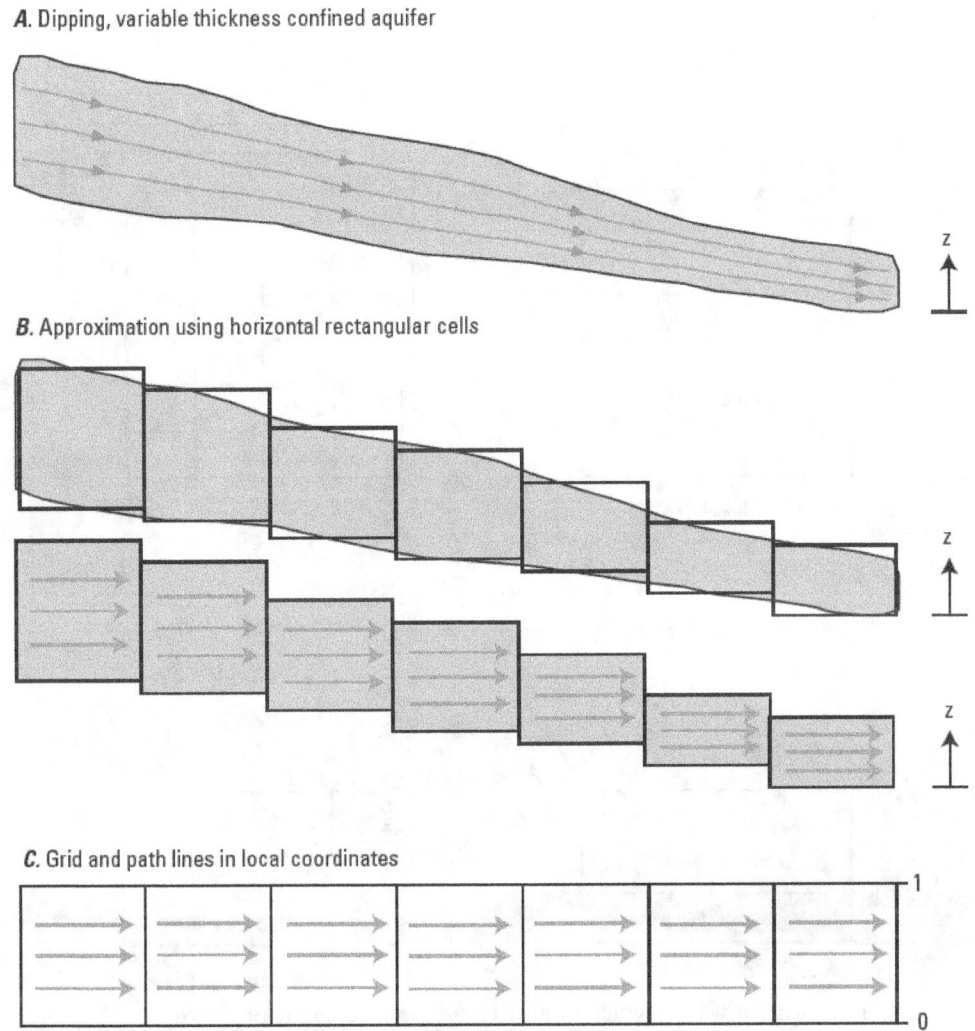

A. Dipping, variable thickness confined aquifer

B. Approximation using horizontal rectangular cells

C. Grid and path lines in local coordinates

Figure 4.   An inclined aquifer with variable thickness.

Quasi Three-Dimensional Confining Beds

Many groundwater systems are characterized by sequences of highly transmissive, sub-horizontal aquifers that are vertically separated by confining beds having much lower transmissivity. Because of the large contrast in hydraulic conductivity between aquifers and confining beds, groundwater flow is predominantly lateral in aquifers and vertical through confining beds. In these systems, confining beds function primarily as low-conductivity vertical connections between aquifer layers. Confining beds sometimes are not simulated as active model layers in such systems. Instead, their effect on vertical flow between aquifers is accounted for implicitly by computing the effective vertical hydraulic conductance between aquifers based on the vertical conductivity and thickness of the confining beds. This approach is referred to as a quasi-three-dimensional representation. In MODPATH, each unsimulated confining bed is assigned to be part of the model layer directly above it. The local z coordinate within the confining bed varies linearly from -1 at the bottom of the confining bed to 0 at the top of the confining bed (fig. 5). It is assumed that one-dimensional, steady-state, vertical flow exists throughout the confining bed. That assumption implies that the average vertical linear velocity is constant throughout the confining bed and that its magnitude equals the volumetric flow rate between adjacent model layers divided by the area of the cell and the porosity of the confining bed. When a particle reaches a top or bottom face of a cell that is a boundary of a confining bed, the particle moves vertically across the confining bed into the next active model layer. Time of travel across the confining bed is computed by dividing the thickness of the confining bed by the average vertical linear velocity within the confining bed. A value for the porosity of the confining bed must be specified to compute travel time across the layer.

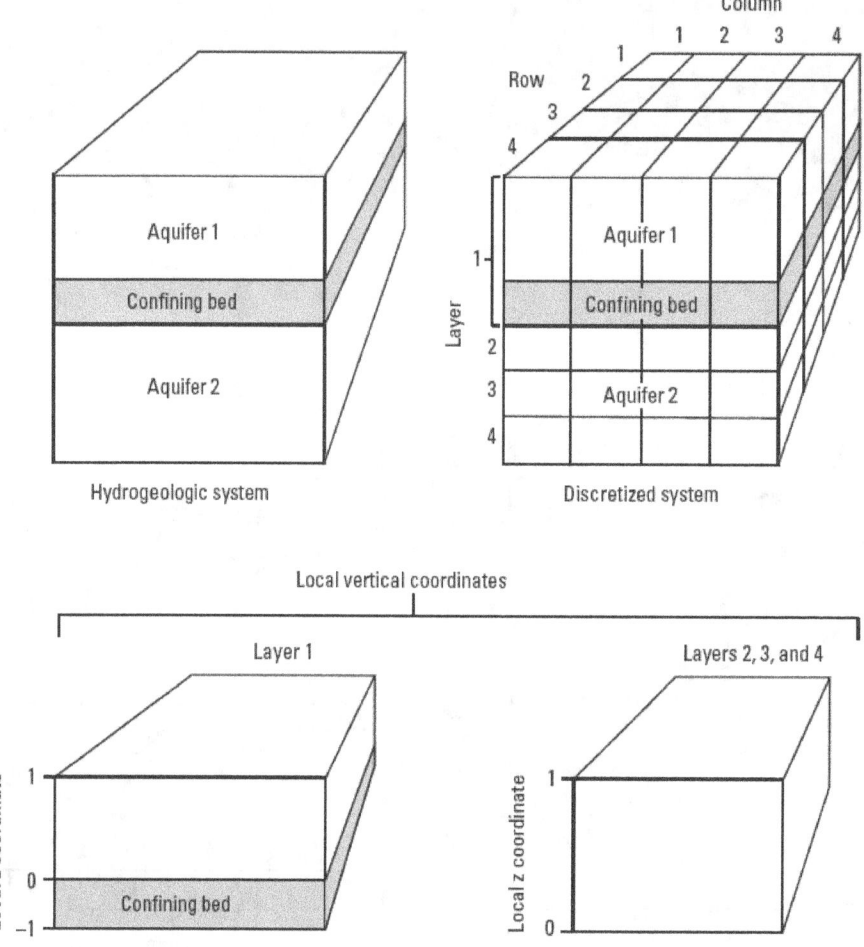

Figure 5.   Definition of local vertical coordinates for quasi-three-dimensional systems.

Backward Tracking

Particles can be tracked backwards in the "upstream" direction by running the particle-tracking algorithm in reverse. Backward tracking is accomplished by reversing the sign of all the velocity components and then applying the algorithm just as described for forward tracking.

Transient Flow

Transient finite-difference flow simulations consist of a series of discrete time steps during which flow rates remain constant and storage changes within cells contribute an additional component to the internal source/sink term on the right side of equation 3. The particle-tracking algorithm described previously for steady-state flow systems can be extended to transient finite-difference simulations by taking advantage of the fact that transient simulations behave as a series of steady-state flow periods. For each time step, particle paths are computed just as for the steady-state case until the end of the time step is reached. A new velocity distribution is then calculated for the next time step and the computation of particle paths is resumed. The computation of paths forward or backward, boundary conditions, and the pathline termination criteria are handled the same as for steady-state flow.

Water-table layers represent an additional complication in transient particle-tracking analyses because the water table is actually a moving boundary. MODPATH deals with this problem in a simple manner by assuming that the water table moves in discrete jumps from one time step to the next. The saturated thickness of a water-table cell is assumed to remain constant over the length of a time step and is computed using the head at the end of the time step. This approach is consistent with the fully-implicit, transient finite-difference scheme employed by MODFLOW in which inter-cell flow rates are computed using saturated thicknesses and hydraulic gradients derived from heads at the end of the time step. MODPATH computes the vertical velocity component at the water table by dividing the volumetric flow rate across the top of the cell by the porosity. For transient flow, this vertical velocity component is actually the vertical velocity component relative to the velocity of the water table.

Consider the case of a falling water table in a situation in which no recharge crosses the water table. A particle placed at the water table would move downward at the same speed as the water table, and no net flow of water would occur across the water-table boundary. At some later time, the particle would still be located at the water table, but its absolute vertical position would be lower. MODPATH accounts for the moving boundary in an approximate way by using the local vertical coordinate of the particle to adjust its absolute vertical coordinate at the beginning of each time step to account for changes in saturated thickness from one time step to the next. Thus, for the example described earlier, MODPATH would hold the particle at the top of the cell as the water table drops from one time step to the next. In cases where there is a non-zero vertical component of flow at the water table, this approach provides a simple way of approximating the vertical movement of particles relative to the water table in a manner consistent with the water balances for unconfined cells.

Solute Retardation

The effect of linear sorption on the apparent velocity of a reactive solute (v^*) can be examined using the particle-tracking method presented in the previous sections by dividing the cell-face velocities in equation 2 by a retardation factor, R:

$$v_x^* = \frac{v_x}{R}, \quad v_y^* = \frac{v_y}{R}, \quad v_z^* = \frac{v_z}{R} \tag{17}$$

For nonreactive solutes that do not interact with the surrounding rock the retardation factor is equal to 1 and the magnitude of the apparent solute velocity is equal to that of the groundwater velocity. For solutes affected by linear sorption, the retardation factor is greater than 1 and the magnitude of the apparent solute velocity is less than that of the groundwater velocity. The retardation factor is a function of the porosity, the density of the geologic material, and the sorption coefficient. A detailed discussion of the retardation factor can be found in Domenico and Schwartz (1990). As implemented by MODPATH, the retardation factor is simply a scaling factor for the magnitude of the velocity that is specified by the user. The retardation factor does not affect the direction of the velocity.

Limitations

MODPATH has a number of limitations that must be understood if it is to be used effectively. These limitations are related to (1) underlying assumptions in the particle-tracking algorithm, (2) discretization effects, and (3) uncertainty in parameters and boundary conditions.

The semianalytical particle-tracking method used in MODPATH is valid only for a linear velocity interpolation scheme. The method is a consistent approach for computing and interpolating velocities from inter-cell flow rates for the standard finite-difference approximation of the groundwater flow equation. MODPATH should work with any of the MODFLOW flow packages that are based on that scheme and that output the standard budget components to the budget output file. A description of the standard budget components that must be present in the budget file for a MODFLOW flow package to be compatible with MODPATH is provided in the *Program Portability and Compatibility* section. MODPATH cannot be used to compute pathlines for other types of numerical approximations of the groundwater flow equation, such as finite-element models or integrated-finite-difference schemes that have more complex cell and node connections.

The accuracy of numerically-generated pathlines, and a proper interpretation of what they represent, depends on the extent to which the groundwater system can be realistically represented by a discrete network of finite-difference cells. The degree of spatial discretization in a finite-difference model influences (1) the level of detail at which hydrogeologic and system boundaries can be represented, (2) the accuracy of velocity calculations, and (3) the ability to accurately and unambiguously represent internal sources and sinks. Often, a level of spatial discretization that is adequate for a flow simulation analysis oriented toward water supply may not be adequate for a pathline analysis. Time discretization also can be a substantial source of error in transient flow simulations.

The effect of spatial discretization on the representation of internal sources and sinks is especially important for particle-tracking analyses because of the ambiguity associated with the movement of particles through weak sink cells. Weak sink cells contain sinks that do not consume water at a large enough rate to capture all of the water that enters the cell. The net result is a

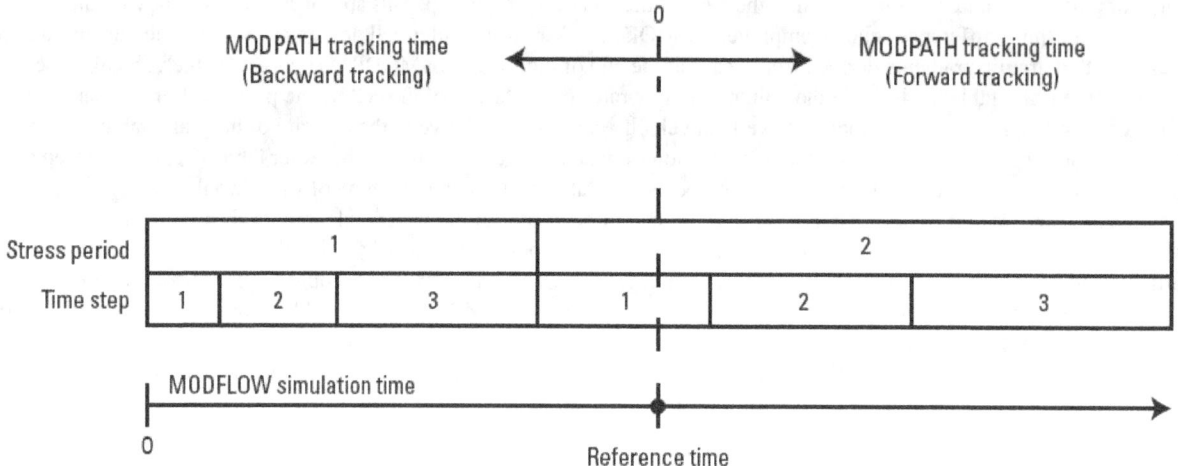

Figure 6.   Time concepts used by MODPATH.

flow-through cell in which some fraction of the total inflow to the cell eventually flows out of the cell across one or more of the cell's faces. Pathlines computed for these cells are consistent with the assumption of a uniformly distributed sink within the cell. It is difficult to interpret the results of particle-tracking analyses in systems with weak sinks for the following reasons:

1.   There is no way to know whether a specific particle should discharge to the sink or pass through the cell. This means that individual particles will not correspond to a fixed volume of water, nor will flow tubes defined by adjacent pathlines represent a fixed quantity of flow.

2.   Pathlines through weak sink cells may not accurately represent the path of any water in the system if the cells contain point sinks that cannot be represented accurately as being uniformly distributed throughout the cells.

These problems are a direct result of spatial discretization that is too coarse. Using a finer grid may eliminate the problem by turning weak sink cells into strong sink cells that clearly correspond to discharge points for particles that enter those cells. In practice, it usually is impossible to entirely eliminate weak sink cells from a field-scale MODFLOW model. Similar issues exist for analyzing recharge locations for particles in backward tracking simulations that have weak source cells.

All of the limitations discussed so far relate to discretization effects and the underlying particle-tracking method. In fact, the most important limitation in any groundwater model is the uncertainty in boundary conditions and hydrogeologic properties used to define the system. At best, a MODPATH particle-tracking analysis only provides information about how water moves within the idealized system represented by the MODFLOW simulation.

MODPATH Overview

The following sections discuss several concepts that are essential to understanding MODPATH's structure and operation.

Time Concepts

The following time concepts are illustrated in figure 6:

- *Simulation Time*—is the value of time associated with the MODFLOW simulation. Simulation time starts at zero and increases throughout the course of a MODFLOW simulation. The range of simulation time is defined by the stress period and time step data in the MODFLOW Discretization (DIS) file.

- *Tracking Time*—is defined for each MODPATH simulation relative to specified reference value of simulation time. Tracking time is defined to be zero at the specified reference simulation time. Tracking time measures the accumulated time during a particle-tracking analysis. The value of tracking time is always positive, regardless of whether particles are tracked in the forward or backward direction.

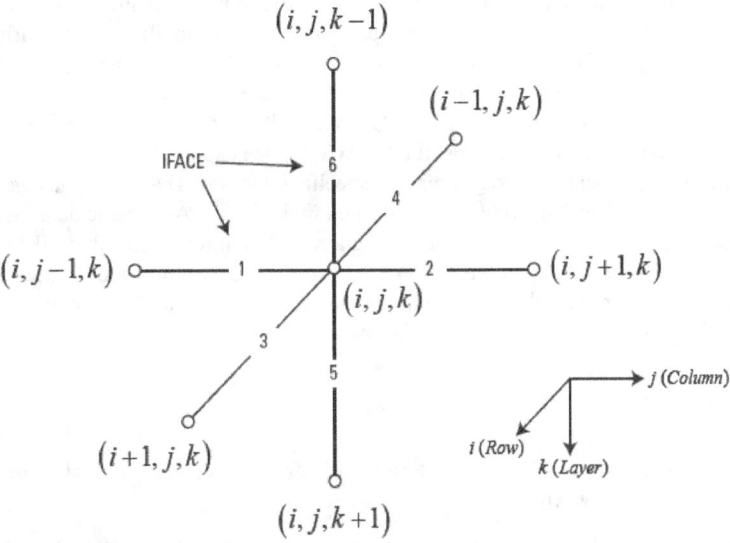

Figure 7.   Definition of the cell face index, IFACE.

The concept of tracking time allows MODPATH to track particles backward in time through either steady-state or transient flow fields simply by reversing the sign of the velocity used in the tracking calculation. That approach greatly simplifies the particle-tracking computation by allowing MODPATH to use the same set of logic and code to simulate the entire range of forward and backward tracking scenarios under both steady-state and transient flow conditions.

Flow Boundaries

It is common for MODFLOW simulations to use stress packages to represent boundary flows. In MODFLOW, stress packages simply add a component of flow to the grid cells in which they are active. In MODFLOW, stress package flows behave as internal sources or sinks of water. In cases where stress package flows are intended to represent flows that enter across system boundaries, however, the velocity field used by MODPATH will more accurately represent the system if those flows are assigned to the appropriate boundary faces of cells and used to compute velocity components across boundary cell faces. A good example is recharge represented with the recharge package. Areal recharge commonly represents inflow to the system across the water table. Consequently, the interpolated velocity field in a water-table cell will be more accurately represented in MODPATH by assigning the recharge flow to the top face of the cell and using it to compute a vertical component of velocity across the top face of the cell. Another common example is the river package, which often is used to represent shallow streams that sit at or near the top of grid cells that represent an unconfined aquifer. As with recharge, the velocity in those cells also will be more accurately represented by treating the river package flows as if they occurred across the top face of the cell. In other cases, such as for the well package, the stress package flows are intended to represent internal sources and sinks. In those cases, no additional information is required by MODPATH because MODPATH automatically represents them as internal, uniformly-distributed sources or sinks. In the case of boundary stress package flows, MODPATH requires additional information indicating which faces are the flow boundary faces.

Boundary face information is provided to MODPATH by assigning each of the six cell faces an index number as illustrated in figure 7. Faces 1 and 2 are perpendicular to the x-direction, faces 3 and 4 are perpendicular to the y-direction, face 5 is the bottom of the cell, and face 6 is the top of the cell. MODPATH refers to cell faces using a variable named IFACE that is assigned to each stress package flow. The value of IFACE indicates the face across which the flow occurs. In the case of true internal sources or sinks, such as wells, the IFACE value is set equal to 0. It is also possible to specify an IFACE value of -1; in such cases, MODPATH distributes the stress package flow for a cell uniformly over all of the lateral faces (faces 1 through 4) that are boundaries of the active model grid. If a cell does not have any lateral boundary faces, the flow rate is treated as an internally distributed sink/source. The option to specify a negative value for IFACE provides a convenient way to represent lateral flow boundaries that are irregularly shaped.

The additional IFACE data are provided to MODPATH in one of two ways, depending on the type of stress package. If a MODFLOW stress package saves IFACE values in the compact budget file as auxiliary data (Harbaugh, 2005), MODPATH reads IFACE values from the compact budget file. However, some stress packages do not support saving IFACE values in the compact budget file. The recharge and evapotranspiration packages are the two most widely used packages that do not support auxiliary IFACE values. For those types of packages, MODPATH allows the user to specify the budget label and a default IFACE value in the MODPATH basic data file. The specified IFACE value is applied to all of the stresses for that package; IFACE information for any number of stress packages may be specified this way. If a stress package that saves IFACE values in the compact budget file also has a default IFACE value specified in the MODPATH basic data file, the IFACE values in the compact budget file are used and the specified default value is ignored. If no information about IFACE is provided in either the compact budget file or the MODPATH basic data file, an IFACE value of 0 is assumed and all stresses are treated as internally distributed sinks or sources. Whenever possible, IFACE values should be saved as auxiliary data in the compact budget file to provide the most flexibility in defining flow boundary conditions.

Terminating Particles

Particles are stopped whenever they reach points of termination or whenever the cumulative tracking time equals the maximum allowed value. A particle terminates when

- it reaches an external boundary face or an internal sink/source cell that captures the particle,
- it enters a cell with a zone code that designates the cell as a termination location,
- the cumulative tracking time has reached the maximum allowed value, or
- it encounters an abnormal condition that will not allow tracking to continue.

MODPATH continues to track particles until all particles have terminated.

Particles always terminate when they reach a cell face that represents a flow boundary for the active grid, as described in the preceding section. Particles also may terminate when they enter a cell containing an internal sink or source. For the forward tracking case, cells that contain internal sinks can be characterized as either strong sink cells or weak sink cells. A strong sink cell is one that contains an internal sink and has inflow from adjacent cells but no outflow to other cells. In contrast, weak sink cells have both inflow and outflow across their cell faces. Their internal sinks remove some of the inflow from adjacent cells, but not all of it. Analogous definitions exist for strong and weak source cells. For weak sink and weak source cells, an arbitrary decision must be made by the user about when to terminate particles. MODPATH provides options to allow particles either to pass through or stop when they enter cells containing weak sinks or weak sources.

Sometimes it is useful to stop particles at an arbitrary location that does not correspond to a natural termination point. MODPATH provides that capability by allowing zone numbers to be assigned to grid cells and identifying a range of zone numbers that indicate particle termination locations. An example of such a situation would be a case where the desire is to determine the recharge locations at the water table that contribute water to a deep aquifer. That could be achieved by assigning a zone number to the cells representing the deep aquifer, indicating they are termination points. Particles that start at the water table would then stop when they reach the deep aquifer because the cells are designated termination locations. The results could be processed to produce a recharge map for the deep aquifer.

Particle tracking is stopped whenever the maximum allowed value for tracking time has been reached. That value may be specified by the user, or it may either be when the cumulative tracking time has reached the time specified for the end of the MODFLOW simulation (if particles are tracked forward in time) or the beginning of the MODFLOW simulation (if particles are tracked backward in time). MODFLOW allows for a mix of steady-state and transient stress periods. MODPATH provides a variety of options for how to handle the situation when a particle-tracking simulation reaches the end or beginning of a MODFLOW simulation. For forward tracking simulations, MODPATH forces the particle-tracking simulation to stop when the end of the MODFLOW simulation is reached and the final stress period is transient. Under those conditions, there is no basis for extending the particle tracking beyond the end of the simulation. If the final stress period is steady state, however, the option is provided to either stop the particle-tracking simulation or extend it indefinitely in time using the velocity distribution of the final time step. If a steady-state time step is extended, particle tracking continues until all particles have terminated. Providing the option to extend the simulation if a steady-state condition exists at the end of the simulation is appropriate because it is reasonable to assume that steady-state flow at the end of the simulation could continue beyond that point in time. The same logic is applied for backward tracking simulations, except in such cases, the decision to stop or extend the particle-tracking simulation is based on whether steady-state or transient flow conditions exist for the first MODFLOW stress period.

Sometimes particles encounter abnormal conditions that require them to be terminated. The most common cause of abnormal termination occurs in transient flow systems where cells are allowed to go dry. Under those conditions, particles are sometimes stranded in cells that become inactive after having gone dry. Particles are terminated in such cases and given a status code that indicates they were stranded

Model Input

MODPATH obtains data from a set of input files that are a combination of MODFLOW input and output files plus additional data files specific to MODPATH. The following sections provide only an overview of the data required by MODPATH. Detailed descriptions of the structure and content of the data files are found in Input Instructions section.

Flow System Files

MODPATH requires the following files that contain the basic information about the groundwater flow system:
- MODPATH basic data file
- MODFLOW discretization file
- MODFLOW head output file
- MODFLOW cell-by-cell compact budget output file

The MODPATH basic data file consists of a combination of data contained in the MODFLOW Basic (BAS) package data file and the MODFLOW flow package data file. It also contains additional data, such as porosity, that is specific to MODPATH and not required by MODFLOW. The MODPATH basic data file may also refer to additional ancillary data files that contain various types of array data.

Information defining spatial and time discretization is obtained directly from the MODFLOW discretization (DIS) file. No additional information is required.

Head output from MODFLOW is used by MODPATH to compute saturated thickness for water-table layers and to provide information about the distribution of dry cells. MODFLOW has the option of saving the head output in either binary or ASCII text format; however, MODPATH requires that the head output be saved as a binary file.

The flow-rate data used by MODPATH to compute groundwater velocity components is obtained from the MODFLOW cell-by-cell budget output. MODPATH requires that the cell-by-cell budget data for all packages be saved in a single file. Although the standard and compact budget output file styles are supported, it is strongly recommended that cell-by-cell budget output be saved using the compact option. Unlike the standard MODFLOW budget file, the compact budget file can contain extra information about stress package flows that allow flow boundary conditions to be represented more accurately by assigning stress package flows to specific cell faces (as described in the *Flow Boundaries* section).

Simulation File

The MODPATH simulation file contains data that define the options for a specific MODPATH simulation, such as the following:
- Output filenames
- Tracking direction
- Simulation type and output options
- Particle starting location data
- Options for terminating particles

It is common to generate several MODPATH simulation files for a single MODFLOW flow simulation in order to examine the groundwater flow from a variety of perspectives. In such instances, the MODPATH flow system files generally will not change from run to run, but the simulation file will change to reflect changes in particle-tracking options.

Particle Starting Data

Starting data for particles can be generated automatically by MODPATH for one or more groups of grid cells specified in the MODPATH simulation file. Starting data also can be input in a variety of ways from a separate starting locations file.

In most instances, the quickest way to generate a large number of particles is to use automatic particle generation. For this procedure, a group of grid cells is defined along with a spatial pattern of particles that will be applied to each cell in the group. The particle pattern may consist of either (1) two-dimensional arrays of particles on one or more cell faces, or (2) a three-dimensional array of particles that distributes particles within a cell. In addition, a *release time* is also specified for the particle group. The release time is a value of tracking time that is measured relative to the reference simulation time. It represents the time at which the particle enters the flow system. An option also is provided to specify a *release period* and the number of additional *release events*. If a release period and additional release events are specified, additional particles are generated at regular time intervals throughout the release period. MODPATH uses the spatial information and the particle release information to generate a complete set of particles for the group. Any number of particle groups can be defined.

Sometimes more control over starting particle data is needed than can be achieved with the automatic generation scheme. In such cases, starting particle data can be input from a separate starting locations data file, the name of which is specified in the simulation file. The starting locations file allows a much larger degree of flexibility in defining both spatial and temporal properties of particles than the starting location options provided in the MODPATH simulation file. The starting locations file also provides options for organizing and optimizing particle starting data (as described in the Input Instructions—Starting Locations File section).

Model Output

The following types of output files are produced by MODPATH:

- *Listing file*—is a summary of model input and results that is generated for each MODPATH simulation. The MODPATH listing file is similar in style to the listing file produced by MODFLOW.

- *Debug files*—are files that provide information about program flow and detailed accounting of particle-tracking calculations that can be helpful for troubleshooting problems when they occur. The debug files include a log file that records the program flow and an optional trace file that can be used to record detailed output on a cell-by-cell basis for a single particle throughout the course of a simulation. The option to generate a trace file is specified in the MODPATH simulation data file.

- *Particle coordinate output files*—are files that record the movement of particles throughout the course of a simulation and form the basis for input data for post-processing programs that analyze and display the results.

Three types of simulations are supported by MODPATH:

- Endpoint simulation

- Pathline simulation

- Timeseries simulation

The actual particle-tracking calculations performed by MODPATH are the same for each of the three simulation types. The only difference between the endpoint, pathline, and timeseries simulations is the style of particle output files that are produced. These output files are summarized next and are described in detail in the section Input Instructions—Particle Coordinate Output Files section.

For an endpoint simulation, only information about the initial and final location and time data are recorded in an endpoint file. This text file is generated for all MODPATH simulations and contains a one-line data record for each particle. Intermediate particle coordinates computed during the simulation are not saved as output. Endpoint simulations produce the minimum amount of output possible for a MODPATH simulation. Endpoint simulations are most useful when the objective is to map recharge locations to discharge locations for a large number of particles.

For pathline simulations, coordinates along the path of each particle are recorded in a pathline file. The pathline file contains the starting coordinates of a particle and the coordinates at every intermediate point computed during the simulation. Pathline files contain all of the information necessary for a graphical post-processing program to display the pathline track. Pathline simulations also produce an endpoint file.

For timeseries simulations, only the locations of particles at specified points in time are recorded in a timeseries file. Timeseries simulations produce a set of particle location records that are stacked in time. When points in a timeseries file are displayed, the progress of a group of particles is shown as a series of snapshots in time. A timeseries simulation can produce an optional advective observations output file that is similar to the timeseries file. As with other simulations, timeseries simulations also produce an endpoint file.

Program Portability and Compatibility

MODPATH is written in FORTRAN and can be compiled by any FORTRAN compiler that fully supports the FORTRAN-95 standard (International Standards Organization, 1997) as well as the STREAM option for reading and writing binary files, which is defined by the FORTRAN-2003 standard (International Standards Organization, 2004). Most FORTRAN compilers available as of 2011 meet those requirements.

MODPATH is fully compatible with MODFLOW-2005 and MODFLOW-2000. Prior to MODFLOW-2000, however, discretization information was not located in a separate discretization file and the auxiliary variable IFACE was not saved in the budget output file. In order to use this version of MODPATH with those earlier versions of MODFLOW, users must create a discretization file to provide input to MODPATH. In addition, versions of MODFLOW prior to MODFLOW-2000 will not be able to take full advantage of IFACE to specify boundary flows in list-oriented stress packages, such as the well package.

The version of MODPATH presented herein requires that the MODFLOW head and budget output files be saved as pure binary files with no extra record delimiters. By default, FORTRAN "unformatted" output commonly contains compiler-dependent record delimiters that will cause MODPATH to read those files incorrectly. To avoid this problem, MODFLOW should be compiled with an appropriate option to write unformatted files without record delimiters. Most FORTRAN compilers have an option to write undelimited, unformatted files. Beginning with FORTRAN-2003, the STREAM option can be used to assure that undelimited, unformatted files are generated.

MODFLOW packages that produce flow output for the budget file can be divided into two major categories: flow packages and stress packages. Flow packages account for the flow of groundwater from one cell to another and for changes in storage that occur under transient conditions. MODFLOW supports a number of flow packages, each with its own customized data requirements and input style. The Layer Property Flow (LPF) package is the most widely used flow package at the present time. Stress packages provide connections to external sources and sinks of water that represent either internally distributed features (such as wells) or flow across spatial boundaries (such as areal recharge). MODPATH interacts with flow packages and stress packages solely through the flow-rate information that each package saves in the budget output file. Because it relies only on the flow information that is present in the budget file, MODPATH is compatible with any flow or stress package that meets certain criteria for saving its flow data in the budget file.

For a MODFLOW flow package to be compatible with MODPATH, it must save the following five flow components in the budget file, in the order listed below, for every time step:

- Storage (only if the stress period is transient)

- Constant head flows

- Flow across the right face of each grid cell (only if the number of columns > 1)

- Flow across the front face of each grid cell (only if the number of rows > 1)

- Flow across the lower face (only if the number of layers > 1)

The header records for the flow components must be labeled with the following 16-character, right-justified text strings:

```
STORAGE
CONSTANT HEAD
FLOW RIGHT FACE
FLOW FRONT FACE
FLOW LOWER FACE
```

Any flow package that adheres to the structure described above will be compatible with MODPATH.

For a MODFLOW stress package to be compatible with MODPATH, it must save its flow-rate information to the budget file. List-oriented stress packages should support auxiliary variables and save their auxiliary variable data to the budget file when the MODFLOW compact budget option is selected. Support for auxiliary variables allows IFACE values to be defined and saved for individual stresses in the stress package list. Although not strictly required by MODPATH, support for auxiliary variables greatly improves the ability to accurately represent flow boundaries. In addition to the requirements just discussed, the MODFLOW budget file should not be used to save any flow rate data that are not a direct component of the MODFLOW groundwater flow budget.

Input Instructions

This section includes input instructions for the MODPATH name file, the MODPATH basic data file, the simulation file, and the starting locations file. Documentation for the MODFLOW discretization file is presented in Harbaugh (2005). Input data items are shown in italics. Data items that correspond directly to MODFLOW data items are represented as all capital letters and have the same name that appears in the MODFLOW documentation. Data items specific to MODPATH are represented by mixed-case names that do not necessarily correspond directly to variable names in the MODPATH source code. Data items that are optional under certain circumstances are enclosed in brackets. The names of data items are shown in bold italics at the location in the text where they are defined and in regular weight italics when they appear elsewhere in the text.

General Structure of Input Data

Free-Format Input

Free-format input is used unless otherwise specified. Free-format style is similar to FORTRAN's list-directed input. With free-format, the spacing of values within a record is not fixed—each value can consist of any number of characters. One or more spaces, or a single comma optionally combined with spaces, must separate adjacent values. A value of zero must be explicitly represented as "0" rather than one or more spaces because there is no way to detect the difference between a space that represents a null value and one that represents a value separator. Two capabilities included in FORTRAN's list-directed input are not included. First, null values are not allowed in an input list. Second, a slash character, "/", cannot be used to terminate an input record; data values for all input items must be explicitly specified. For character data, MODPATH's free-format implementation is less stringent that FORTRAN's list-directed input. FORTRAN requires character data to be delineated by apostrophes. MODPATH does not require apostrophes if the character data item contains no spaces. Furthermore, if the character data item is the only data item on a line, apostrophes are not required even if the character data item includes spaces. If multiple data items appear on a single line, any character data items that contain spaces must be delineated by apostrophes.

Array Input

MODPATH uses the same array reader subroutines (U2DREL, U2DINT, and U1DREL) used by MODFLOW (Harbaugh, 2005). Detailed documentation of array input can be found in Harbaugh (2005). For convenience, a brief summary of the five array control record styles is provided below.

FIXED-FORMAT CONTROL RECORD FOR REAL AND INTEGER ARRAY READERS:

Item 1   LOCAT   CNSTNT   FMTIN   IPRN
           Format: I10, F10.0, A20, I20

FREE-FORMAT CONTROL RECORD FOR REAL ARRAY READERS

For each of the control record options, the first data item (shown in bold) is a key word that can be specified as uppercase or lowercase. Each control record is limited to a length of 79 characters.

Option 1.   **CONSTANT**   CNSTNT
            All values in the array are set equal to the value CNSTNT.

Option 2.   **INTERNAL**   CNSTNT   FMTIN   IPRN
            The individual array elements will be read from the same file that contains the control record.

Option 3.   **EXTERNAL**   Nunit   CNSTNT   FMTIN   IPRN
            The individual array elements will be read from the file unit number specified by Nunit. The name of the file associated with this file unit is contained in the name file.

Option 4.   **OPEN/CLOSE**   FNAME   CNSTNT   FMTIN   IPRN
            The array will be read from the file whose name is specified by FNAME. This file will be opened on file unit 99 just prior to reading the array and closed immediately after the array is read. Thus, a file that is read using this control record can contain only a single array. Files opened using the OPEN/CLOSE option should not be included in the name file.

Explanation of Variables used in the Array-Control Records

Nunit—is the unit for reading the array when the EXTERNAL free-format control record is used.

CNSTNT—is a real-number constant for U2DREL and U1DREL, and an integer constant for U2DINT. If the array is being defined as a constant, CNSTNT is the constant value. If individual elements of the array are being read, the values are multiplied by CNSTNT after they are read. When CNSTNT is used as a multiplier and iis specified as 0, it is changed to 1.

FMTIN—is the format for reading array elements. The format must contain 20 characters or less. The format must either be (1) a standard FORTRAN format that is enclosed in parentheses, (2) "(FREE)", which indicates free-format, or (3) "(BINARY)", which indicates binary data. When using a free-format control record, the format must be enclosed in apostrophes if it contains one or more blanks or commas. The binary files that can be read by the array readers must have either been created by MODFLOW or by another program capable of generating binary files with the appropriate structure. (FREE) and (BINARY) can only be specified in free-format control records. Also, (BINARY) can only be specified when using U2DREL or U2DINT, and only when the control record is EXTERNAL or OPEN/CLOSE.

LOCAT—indicates the location of the array values for a fixed-format array control record. If *LOCAT* = 0, all elements are set equal to CNSTNT. If *LOCAT* > 0, it is the unit number for reading formatted records using FMTIN as the format. If *LOCAT* < 0, it is the unit number for binary records, and *FMTIN* is ignored. In addition, when *LOCAT* is not 0, the array values are multiplied by *CNSTNT* after they are read.

IPRN—is a flag indicating that the array being read should be printed and a code for indicating the format that should be used. An *IPRN* value of 0 will produce readable output for all real and integer data. If *IPRN* is set to a negative number, the array will not be printed. Additional print codes can be specified to customize the format of the printed arrays. The available values of *IPRN* are documented in Harbaugh (2005).

Simulation File

The name of the Simulation File may be specified on the MODPATH command line when the program is executed. If no filename is provided on the command line, MODPATH will prompt the user to enter the name of the simulation file.

The simulation file requires that any data array read using MODFLOW-style array control record use either the CONSTANT, INTERNAL, or OPEN/CLOSE keywords. The EXTERNAL keyword free-format control record and the fixed-format array control record are not supported. Restricting array data input to CONSTANT, INTERNAL, or OPEN/CLOSE styles greatly simplifies the task of managing and editing large numbers of simulation files that reference the same MODPATH Name file.

Data Items

Comments

Item 0   [#*Text*]
          Item 0 is optional. The # symbol must be in column 1. Item 0 can be repeated multiple times.

Filename and Listing Files

Item 1   *ModpathNameFile*

Item 2   *ModpathListingFile*

Option Flags

Item 3   *SimulationType TrackingDirection WeakSinkOption WeakSourceOption, RefercemceTimeOption StopOption ParticleGenerationOption TimePointOption BudgetOutputOption ZoneArrayOption RetardationOption AdvectiveObservationsOption*

          Option flags are listed on a single line and are separated by one or more spaces or a comma

Particle Output Filenames

Item 4   *EndpointFile*

Item 5   [*PathlineFile*] — only if *SimulationType* = 2

Item 6   [*TimeseriesFile*] — only if *SimulationType* = 3

Item 7   [*AdvectiveObservationsFile*] — only if *AdvectiveObservationsOption* = 2 and *SimulationType* = 3

Reference Time

Item 8   [*ReferenceTime*] — only if *ReferenceTimeOption* = 1

Item 9   [*Period Step TimeFraction*] — only if *ReferenceTimeOption* = 2

Stopping Time

Item 10   [*StopTime*] — only if *StopOption* = 3

Particle Starting Locations

If *ParticleGenerationOption* = 1 read automatic particle generation data from this file. Include data items 11 through 21 as appropriate. If *ParticleGenerationOption* =2, skip to data item 22.

Item 11   [*GroupCount*]

For each particle group (1 through *GroupCount*), read data items 12 through 21 as appropriate.

Item 12   [*GroupName*]

Item 13   [*Grid GridCellRegionOption PlacementOption ReleaseStartTime ReleaseOption CHeadOption*]

Item 14   [*ReleasePeriodLength ReleaseEventCount*] — only if *ReleaseOption* = 2

Item 15   [*MinLayer MinRow MinColumn MaxLayer MaxRow MaxColumn*] — only if *GridCellRegionOption* = 1

Item 16   [*Mask(NCOL,NROW)*] — U2DINT — only if *GridCellRegionOption* = 2; an array is read for each layer in the grid.

Item 17   [*MaskLayer*] — only if *GridCellRegionOption* = 3

Item 18   [*Mask(NCOL,NROW)*] — U2DINT – only if *GridCellRegionOption* = 3; A single *Mask* array is read. The layer to which it applies is *MaskLayer* specified in item 17.

Item 19   [*FaceCount*] — only if *PlacementOption* = 1

Item 20   [*IFace ParticleRowCount ParticleColumnCount*] — only if *PlacementOption* = 1. Repeat data item 20 *FaceCount* times. *IFace* must have a value from 1 to 6 for this data item.

Item 21   [*ParticleLayerCount ParticleRowCount ParticleColumnCount*] — only if *PlacementOption* = 2

If *ParticleGenerationOption* = 2, read particle data from a starting locations file.

Item 22.   [*StartingLocationsFile*] — only if *ParticleGenerationOption* = 2

Specified Time Points

If *ParticleGenerationOption* = 1, skip data items 23 through 25.

Item 23   [*TimePointCount*] — only if *TimePointOption* = 2 or *TimePointOption* = 3

Item 24   [*ReleaseTimeIncrement*] — only if *TimePointOption* =2

Item 25   [*TimePoints(TimePointCount)*] — only if *TimePointOption* = 3

Budget Output

If *BudgetOutputOption* = 1 or 2, skip data items 26 through 29.

Item 26   [*CellBudgetCount*] – only if *BudgetOutputOption* = 3

Item 27   [*Grid Layer Row Column*] – only if *BudgetOutputOption* = 3.
Repeat item 27 *CellBudgetCount* times.

Item 28   [*TraceFile*] – only if *BudgetOutputOption* = 4

Item 29   [TraceID] – only if BudgetOutputOption = 4

Zones

If *ZoneArrayOption* = 1, skip data items 30 and 31.

Item 30   [StopZone]

For each layer in the grid read:

Item 31   [Zone(NCOL,NROW)] –U2DINT

Retardation Factor

If *RetardationOption* =1, skip data items 32 and 33.

For each layer in the grid read:

Item 32   [RetardationFactor(NCOL,NROW)] –U2DREL

Item 33   [RetardationFactorCB(NCOL,NROW)] –U2DREL – only if LAYCBD in the MODFLOW Discretization File is not equal to 0

Explanation of Variables

NCOL, NROW, NLAY—is the number of columns, rows, and layers in a MODFLOW grid.

Text—is a character variable (199 characters) that starts in column 2. Any characters can be included in *Text*. The "#" character must be in column 1. Lines beginning with # are restricted to the first lines of the file. Text is written to the Listing File.

ModpathNameFile—is the name of the MODPATH Name File. The filename may be up to 200 characters long. The name should be a local filename, not a full pathname. Spaces are not allowed.

ModpathListingFile—is the name of the MODPATH Listing File. The filename may be up to 200 characters long. The name should be a local filename, not a full pathname. Spaces are not allowed.

SimulationType—is an integer indicating the type of particle-tracking simulation.

1 = Endpoint simulation

2 = Pathline simulation

3 = Timeseries simulation

TrackingDirection—is an integer indicating the direction of the particle-tracking computation.

1 = Forward tracking

2 = Backward tracking

WeakSinkOption—is an integer indicating how weak sinks are treated.

1 =   Allow particles to pass through cells that contain weak sinks.

2 =   Stop particles when they enter cells that contain weak sinks.

WeakSourceOption—is an integer indicating how weak sources are treated.

1 =   Allow particles to pass through cells that contain weak sources.

2 =   Stop particles when they enter cells that contain weak sources.

ReferenceTimeOption—is an integer indicating how the MODPATH reference time will be specified.

1 =   Specify a value for reference time.

2 =   Specify a stress period, time step, and relative time position within the time step to use to compute the reference time.

StopOption—is an integer indicating how the particle-tracking simulation should be terminated.

1 =   For forward tracking simulations, stop at the end of the MODFLOW simulation. For backward tracking simulations, stop at the beginning of the MODFLOW simulation.

2 =   Extend the initial or final steady-state MODFLOW time step as far as necessary to track all particles through to their termination points.

For forward tracking simulations, this option would have an effect whenever the final MODFLOW stress period is steady-state. For backward tracking simulations, this option would have an effect whenever the first MODFLOW stress period is steady-state. If all MODFLOW stress periods are transient, option 2 produces the same result as option 1.

3 =   Specify a value of tracking time at which to stop the particle-tracking computation.

ParticleGenerationOption—is an integer indicating how particle starting locations are generated.

1 =   Specify information to automatically generate particles for a collection of cells.

2 –   Read particle locations from a starting locations file.

TimePointOption—is an integer indicating if particle locations are computed at specified time points.

1 =   Time points are not specified.

2 =   A specified number of time points are calculated for a fixed time increment.

3 =   An array of time point values is specified.

TimePointOption must be set to 1 for endpoint simulations (*SimulationType* = 1). *TimePointOption* must be set to 2 or 3 for timeseries simulations (*SimulationType* = 3).

BudgetOutputOption—an integer indicating the budget checking option.

1 =   No budget checking

2 =   A summary of cell-by-cell budgets is printed in the Listing File

3 =   A list of cells is specified for which detailed budget information is summarized in the Listing File

4 =   Trace mode is in effect

For trace mode, detailed budget and cell information is output for a single particle specified by the user. Trace data are written to a separate output file.

ZoneArrayOption—is an integer indicating if a zone array will be read.

> 1 = No zone data are read.

> 2 = Zone data are read.

If option 1 is selected, MODPATH sets the zone value for every cell equal to 1.

RetardationOption—is an integer indicating if a retardation factor array will be read.

> 1 = Retardataion factors are not read or used in the velocity calculations.

> 2 = An array of retardation factors is read and used in the velocity calculations.

AdvectiveObservationsOption—is an integer indicating if advective observations are computed and saved as output.

> 1 = Advective observations are not computed or saved.

> 2 = Advective observations are computed and saved for all time points.

> 3 = Advective observations are computed and saved only for the final time point.

EndpointFile—is the name of the Endpoint File. The filename may be up to 200 characters long. The name should be a local filename, not a full pathname. Spaces are not allowed.

PathlineFile—is the name of the pathline file. The filename may be up to 200 characters long. The name should be a local filename, not a full pathname. Spaces are not allowed.

TimeseriesFile—is the name of the Timeseries File. The filename may be up to 200 characters long. The name should be a local filename, not a full pathname. Spaces are not allowed.

AdvectiveObservationsFile—is the name of the Advective Observations File. The filename may be up to 200 characters long. The name should be a local filename, not a full pathname. Spaces are not allowed.

ReferenceTime—is the reference time for particle-tracking computations. The reference time is a value of MODFLOW simulation time. It represents the starting time for the particle-tracking simulation. Tracking time in a MODPATH simulation is measured relative to the reference time. The reference time must be greater than or equal to 0 and less than or equal to the cumulative time at the end of the last MODFLOW time step.

Period, *Step*, *TimeFraction*—are the stress period, time step, and relative position within the time step used to compute *ReferenceTime* when *StopOption* = 2. *TimeFraction* is a value between 0 and 1, where 0 corresponds to the beginning of the time step and 1 corresponds to the end of the time step.

StopTime—is a specified value of tracking time at which to stop the particle-tracking computation. *StopTime* is always greater than or equal to 0.

GroupCount—is the number of particle groups. When specified, *GroupCount* should be > 0.

GroupName—is the name of the particle group. *GroupName* has a maximum length of 16 characters. Spaces are not allowed.

Grid—is an integer indicating the grid number. This version of MODPATH only supports a single grid; therefore, *Grid* must be set equal to 1.

GridCellRegionOption—is an integer indicating the option for defining the collection of cells in a particle group when automatic particle generation is used (*ParticleGenerationOption* = 1).

> 1 = A 3D rectangular block of cells is specified.

2 =   A 3D mask array is read to define the cells in the group. The cells in the group are defined by those cells having a mask value greater than 0.

3 =   A single layer mask array is read. The mask values of the layer array correspond to the model layer corresponding to the value of MaskLayer read in data item 17.

PlacementOption—is an integer indicating the option for distributing particles within cells during the autogeneration process.

1 =   Specify a 2D array of particles on one or more cell faces of each cell in the particle group.

2 =   Specifiy a 3D array of particles internally within each cell in the particle group.

ReleaseStartTime—is the release time for starting particles in a particle group. *ReleaseStartTime* is a value of tracking time; therefore, it is always greater than or equal to 0. Values of tracking time are measured relative to the reference time of the particle-tracking simulation. If the multiple release option is selected for a group (*ReleaseOption* = 2), *ReleaseStartTime* represents the initial time at which particles in that group are released.

ReleaseOption—is an integer indicating whether a single or multiple release mode will be used for particles in a particle group.

1 =   A single release time will be used for all particles in a group.

2 =   Particles in a group will be released over a period of time beginning at *ReleaseStartTime* and continuing for a specified length of time defined by the length of the release period (*ReleasePeriodLength*) and the number of particle releases (*ReleaseEventCount*).

3 =   Particles in a group are released at specified time values. The additional particle releases are defined by specifying the number of additional releases (*ReleaseEventCount*) and an array of time values (*MultipleReleaseTimes*).

MultipleReleaseTimes—is an array of tracking time values that defines when multiple particle releases occur if ReleaseOption = 3. Tracking time values are measured relative to the reference time of the simulation and are always greater than or equal to zero.

CHeadOption—is an integer indicating whether particles are generated for constant head cells.

1 =   Particles are not generated for constant head cells.

2 =   Particles are generated for constant head cells.

ReleasePeriodLength—is the length of the release period for a particle group that has multiple particle release times. *ReleasePeriodLenth* should be set to a value > 0.

ReleaseEventCount—is the number of additional particle release events for each particle location in a particle group when the automatic particle generation option (*ParticleGenerationOption* = 1) and the multiple release option (*ReleaseOption* = 2) are selected. *ReleaseEventCount* must be > 0. A multiple release of particles consists of (ReleaseEventCount + 1) particle releases starting at time *ReleaseStartTime* and ending at time (*ReleaseStartTime* + *ReleasePeriodLength*).

MinLayer, *MinRow*, *MinColumn*, *MaxLayer*, *MaxRow*, *MaxColumn*—are the grid cell indices that define the three-dimensional block of cells for which particles are generated when automatic particle generation option is selected and *GridCellRegionOption* = 1.

Mask(NCOL,NROW)—is an integer array that defines the distribution of grid cells used to compute particle locations for a particle group when automatic particle generation option is selected and *GridCellRegionOption* = 2 or 3. Particles are generated only for those cells that have a *Mask* value > 0. When *GridCellRegionOption* equals 2, a 2D mask array is read for each model layer. When *GridCellRegionOption* equals 3, only a single 2-dimensional mask array is read.

MaskLayer—is a specified layer used to define grid cells in a particle group when automatic particle generation is used and *GridCellRegionOption* = 3. When *MaskLayer* = 0, particles are placed in the uppermost active model layer at each areal grid-cell location at the specified release time for each particle. If no active model layer exists at a specified particle release time, the particle is ignored.

FaceCount—is the number of cell faces that will have particles assigned to them by automatic particle generation when *PlacementOption* = 1. *FaceCount* has a value in the range 1 to 6.

IFace—is an index number used to associate a flow term with a grid cell face. *IFace* values between 1 and 6 correspond to the six grid cell faces as defined previously in this report. An *IFace* value of 0 is used in certain situations to denote a flow that is assumed to be distributed internally within a cell. An *IFace* value of -1 is used in certain situations to indicate that flow should be distributed uniformly over boundary faces that are perpendicular to the lateral flow direction (faces 1 through 4).

ParticleLayerCount, *ParticleRowCount*, *ParticleColumnCount*—are the number of layer, row, and column subdivisions used to define arrays of particles within cells or on cell faces when automatic particle generation is used.

In the case of a 3D array of particles, *ParticleLayerCount* is the number of vertical subdivisions, *ParticleRowCount* is the number of subdivisions in the y-direction, and *ParticleColumnCount* is the number of subdivisions in the x-direction.

In the case of a 2D array of particles on a cell face, the following definitions apply: For faces 1 through 4, *ParticleRowCount* is the number of vertical subdivisions and *ParticleColumnCount* is the number subdivisions in the horizontal direction (the y-direction for faces 1 and 2; the x-direction for faces 3 and 4). For faces 5 and 6, *ParticleRowCount* is the number of subdivisions in the y-direction and *ParticleColumnCount* is the number of subdivisions in the x-direction.

StartingLocationsFile—is the name of the Starting Locations File. The filename may be up to 200 characters long. The name should be a local filename, not a full pathname. Spaces are not allowed.

TimePointCount—is the number of specified time points used in a pathline or timeseries simulation.

ReleaseTimeIncrement—is the time increment used to compute specified time points when *TimePointOption* = 2.

TimePoints(*TimePointCount*)—is an array of specified time points that is read when *TimePointOption* = 3.

CellBudgetCount—is the number of grid cells for which detailed budget information will be generated.

Grid , *Layer*, *Row*, *Column*—are the grid number and grid cell indices of a cell for which detailed budget information will be generated.

TraceFile—is the name of the Trace file. The filename may be up to 200 characters long and may include spaces.

TraceID—identifies the particle for which trace data will be generated when *BudgetOutputOption* = 4

StopZone—is a zone number indicating that particles should be terminated when they enter cells with that zone value. Set *StopZone* = 0 to indicate that particles should not be stopped based on a zone number of a grid cell. MODPATH automatically sets *StopZone* = 0 when *ZoneArrayOption* = 1.

Zone(NCOL,NROW)—is an integer array of zone numbers. IZONE values should be > 0.

RetardationFactor(NCOL,NROW) —is an array of retardation factors.

RetardationFactorCB(NCOL,NROW) —is an array of retardation factors for a quasi-3D confining bed.

Basic Data File

The MODPATH Basic Data (MPBAS) File contains basic data about the MODFLOW simulation. The name of the MPBAS File is provided in the MODPATH Name File.

Data Items

Item 0   [#*Text*]
                Item 0 is optional. The # symbol must be in column 1. Item 0 can be repeated multiple times. The lines of text are printed in the MODPATH Listing File.

Item 1   *HNOFLO* HDRY

Item 2   *DefaultIFaceCount*

Repeat data items 3 and 4 *DefaultIFaceCount* times.

Item 3   BudgetLabel

Item 4   *DefaultIFACE*

Item 5   *LAYTYP(NLAY)*

Item 6   *IBOUND(NCOL,NROW)*—U2DINT. An array is read for each layer in the grid.

For each layer in the grid, read data items 7 and 8.

Item 7   *Porosity(NCOL,NROW)*—U2DREL

Item 8   *PorosityCB(NCOL,NROW)*—U2DREL, only when the value of *LAYCBD* in the MODFLOW discretization file is > 0 for the layer, indicating the presence of a quasi-3D confining bed. The variable LAYCBD is defined in the MODFLOW discretization file.

Explanation of Variables

Text—is a character variable (199 characters) that starts in column 2. Any characters can be included in *Text*. The "#" character must be in column 1. Lines beginning with # are restricted to the first lines of the file. *Text* is written to the Listing File.

HNOFLO—is the value of head assigned to inactive cells by MODFLOW.

HDRY—is the value of head assigned to cells that are converted to the dry state during a MODFLOW simulation.

DefaultIFaceCount—is an integer specifying the number of budget items for which a default value of the boundary face option is specified (*DefaultIFACE*).

BudgetLabel—is the text label used in the MODFLOW budget file to label the flow rates for a specific budget item. The budget labels are defined in the MODFLOW documentation for each stress package. Budget labels in the MODFLOW budget file are 16 characters long. MODPATH strips leading and trailing spaces from the budget label, however, so budget labels that are right-justified in the MODFLOW budget file do not need to be right-justified in the MODPATH input.

DefaultIFace—is an integer that specifies the default value of *IFace* that will be applied to all cells for the specified budget item. *IFace* indicates how the flow rate for the specified MODFLOW budget item is treated by MODPATH. *DefaultIFace* may have values from -1 through 6:

     -1 =   Flow is uniformly distributed on the lateral cell faces (1 through 4) that correspond to boundaries of the active grid. If no lateral boundary faces are found for a cell, the flow rate is treated as an internally distributed sink/ source.

      0 =   Flow is treated as an internally distributed sink/source.

   1 – 6 =   Flow is assigned to the cell face corresponding to the *IFace* code (1 through 6).

LAYTYP—is an integer flag indicating the MODFLOW layer type.

   0 =   Confined

   1 =   Fully convertible

IBOUND—is the MODFLOW boundary array.

Porosity—is the porosity of a grid layer.

PorosityCB—is the porosity of a quasi-3D confining bed.

Name File

The MODPATH Name File contains the names of several input files used by MODPATH. The Name File is constructed as follows:

Data items

Item 1. *FileType FileUnit FileName FileStatus*
The Name File contains one of the above lines (item 1) for each file. All variables are free format. The length of each line must be 299 characters or less. The lines can be in any order.
Comment lines are indicated by a # character in column 1. Comment lines can appear anywhere in the file and are printed in the MODPATH Listing File.

Explanation of Variables

FileType—is the file type, and it must be one of the following character values. *FileType* can be entered in any combination of upper and lower case.

    MPBAS for the MODPATH Basic Data File

    DIS for the MODFLOW Discretization File

    HEAD for the MODFLOW binary head output file

    BUDGET for the MODFLOW compact budget output file

    DATA for formatted text files

    DATA(BINARY) for unformatted (binary) data files

FileUnit—is the Fortran unit to be used when reading from or writing to the file. Any legal unit number on the computer being used can be specified except units from 80 to 99, which are reserved for internal use by MODPATH.

FileName—is the name of the file, which is a character value. *FileName* may be a local name or a fully-qualified pathname and may be up to 200 characters long. If the name includes spaces, it must be enclosed by apostrophes. Apostrophes are not needed if the name does not include spaces.

FileStatus—is the optional file status, which applies only to file types DATA and DATA(BINARY). Two values are allowed: OLD and REPLACE. OLD indicates that the file should already exist. REPLACE indicates that if the file already exists, it should be deleted before opening a new file. The default is to open the existing file if the file exists or create a new file if the file does not exist.

Starting Locations File

Particle starting location data are read from a separate starting locations file when option flag *ParticleGenerationOption* equals 2 in the MODPATH simulation file. Three input styles are supported:

Input style 1.   The location, release time, and particle ID are specified explicitly for each particle in the simulation. This style provides the flexibility to generate starting locations based on particle output from a previous MODPATH simulation and retain the same particle ID values from one simulation to the next.

Input style 2.   Spatial locations are specified explicitly for one or more particle groups. Particle release time information is specified for each particle group and applied to all particles in a group. Both single and multiple particle release times are supported. MODPATH generates the required particles at the specified locations and release times, and particle ID values are generated automatically. Style 2 allows complete flexibility in specifying the spatial location of particles but retains the convenience and compactness of specifying release time characteristics by particle group. Input style 2 produces a smaller data file size than input style 1.

Input style 3.   Particle starting locations are generated for user-specified regions of grid cells based on a specified template of particle locations located on either cell faces or within the cell for all cells in the specified regions. Both single and multiple particle release times are supported. MODPATH generates the required particles at the specified locations and release times, and particle ID values are generated automatically. Style 3 is equivalent to the format used to specify starting locations directly in the MODPATH simulation file. Input style 3 produces a much smaller data file than either style 1 or style 2.

Data Items

Item 0   *[#Text]*
Item 0 is optional. The # symbol must be in column 1. Item 0 can be repeated multiple times.

Item 1   *InputStyle*

If *InputStyle* = 1 include data items 2 through 5.

Item 2   *GroupCount*

Include data items 3 and 4 for each particle group (1 through *GroupCount*).

Item 3   *GroupName*

Item 4   *GroupParticleCount*
The total number of particles in the simulation is the sum of the number of particles all of the groups.

Include data item 5 for each particle in the simulation.

Item 5   *ParticleID, GroupNumber, Grid, Layer, Row, Column, LocalX, LocalY, LocalZ, ReleaseTime Label*

ParticleID values may be any positive integer and are not required to be a continuous sequence of numbers. Data item 5 records must be read in the order of ascending values of *ParticleID*. MODPATH requires a unique value of *ParticleID* for each particle in the simulation.

When the specified model layer equals 0, particles are placed in the top-most active model layer at each areal grid cell location at the specified release time for each particle. If no active model layer exists at a specified particle release time, the particle is ignored.

If *InputStyle* = 2 include data items 6 through 12.

Item 6   *GroupCount*

For each particle group (1 through *GroupCount*), include data items 7 through 12.

Item 7   *GroupName*

Item 8   *LocationCount, ReleaseStartTime, ReleaseOption*

Item 9   [*ReleaseEventCount, ReleasePeriodLength*] — only if *ReleaseOption* = 2

Item 10   [*ReleaseEventCount*] — only if *ReleaseOption* = 3

Item 11   [*MultipleReleaseTimes(ReleaseEventCount)*] — only if *ReleaseOption* = 3
              Data are provided as a one-dimensional array of time values. Multiple values may be placed on a line.
              As many lines as necessary will be read.

Repeat data item 12 for each particle location in the group (a total of *LocationCount* lines).

Item 12   *Grid, Layer, Row, Column, LocalX, LocalY, LocalZ, Label*

              When the specified model layer equals 0, particles are placed in the top-most active model layer at
              each areal grid cell location at the specified release time for each particle. If no active model layer ex-
              ists at a specified particle release time, the particle is ignored.

If *InputStyle* = 3, include data items 13 through 23.

Item 13   *GroupCount*

For each particle group (1 through GroupCount), read data items 12 through 21 as appropriate.

Item 14   *GroupName*

Item 15   *Grid GridCellRegionOption PlacementOption ReleaseStartTime ReleaseOption CHeadOption*

Item 16   [*ReleasePeriodLength ReleaseEventCount*] — only if *ReleaseOption* = 2

Item 17   [*ReleaseEventCount*] — only if *ReleaseOption* = 3

Item 18   [*MultipleReleaseTimes(ReleaseEventCount)*] — only if *ReleaseOption* = 3
              Data are provided as a one-dimensional array of time values. Multiple values may be placed on a line.
              As many lines as necessary will be read.

Item 19   [*MinLayer MinRow MinColumn MaxLayer MaxRow MaxColumn*] — only if
              GridCellRegionOption = 1

Item 20   [*Mask(NCOL,NROW)*] — U2DINT — only if *GridCellRegionOption* = 2; an array is read for each
              layer in the grid

Item 21   [*MaskLayer*] — only if *GridCellRegionOption* = 3

Item 22   [*Mask(NCOL,NROW)*] — U2DINT — only if *GridCellRegionOption* = 3; a single *Mask* array is
              read that corresponds to the value of *MaskLayer* specified in data item 17.

Item 23   [*ParticleLayerCount ParticleRowCount ParticleColumnCount*] — only if *PlacementOption* = 2

Explanation of Variables

Text—is a character variable (199 characters) that starts in column 2. Any characters can be included in *Text*. The "#" character
must be in column 1. Lines beginning with # are restricted to the first lines of the file.

InputStyle—an integer indicating the style of the starting location data.

    1 = Explicit starting location and time for each particle

    2 = Semiautomatic starting location generation

    3 = Fully automatic starting location generation

GroupCount—is the number of particle groups.

GroupName—is a text string identifying the group. *GroupName* has a maximum length of 16 characters. Spaces are not allowed.

GroupParticleCount—is the number of particles in a group.

ParticleID—is a unique integer used to identify a particle.

Grid—is an integer indicating the grid number for the current particle location. This version of MODPATH only supports a single grid; therefore, *Grid* must equal 1.

Layer—is the layer index for the current particle location.

Row—is the row index for the current particle location.

Column—is the column index for the current particle location.

LocalX—is the local x-coordinate value for the current particle location. The *LocalX* direction is oriented along a grid row and increases from 0 to 1 within a cell in the direction of increasing column index.

LocalY—is the local y-coordinate value for the current particle location. The *LocalY* direction is oriented parallel to grid columns and increases from 0 to 1 within a cell in the direction of decreasing row index.

LocalZ—is the local z-coordinate value for the current particle location. The *LocalZ* direction is oriented vertically and increases from 0 at the bottom of the cell to 1 at the top of the cell. If a layer has an associated quasi-3D confining bed, *LocalZ* varies from -1 at the bottom of the confining bed to 0 at the top of the confining bed (the bottom of the grid cell).

ReleaseTime—the value of tracking time at which the particle is released.

LocationCount—the number of particle locations specified for a particle group for input style 2.

ReleaseStartTime—the initial value of tracking time at which particles in a particle group are released for *InputStyle* = 2.

ReleaseOption—is an integer indicating whether a single or multiple release mode will be used for particles in a particle group.

   1 =   A single release time will be used for all particles in a group

   2 =   Particles in a group will be released over a period of time beginning at *ReleaseStartTime* and continuing for a specified length of time defined by the length of the release period (*ReleasePeriodLength*) and the number of particle releases (*ReleaseEventCount*)

   3 =   Particles in a group are released at specified time values. The additional particle releases are defined by specifying the number of additional releases (*ReleaseEventCount*) and an array of time values (*MultipleReleaseTimes*).

ReleaseEventCount—the number of particle release events for input style 2 when *ReleaseOption* = 2.

ReleasePeriodLength—the total length of time over which particles are released when the multiple release option is used (*ReleaseOption* = 2).

MultipleReleaseTimes—is an array of tracking time values that defines when multiple particle releases occur if ReleaseOption = 3. Tracking time values are measured relative to the reference time of the simulation and are always greater than or equal to zero.

Label—is a user-specified text string of up to 40 characters. Spaces are not permitted.

MinLayer, *MinRow*, *MinColumn*, *MaxLayer*, *MaxRow*, *MaxColumn*—are the grid-cell indices that define the three-dimensional block of cells for which particles are generated when automatic particle generation option is selected and *GridCellRegionOption* = 1.

Mask(NCOL,NROW)—is an integer array that defines the distribution of grid cells used to compute particle locations for a particle group when the automatic particle generation option is selected and *GridCellRegionOption* = 2 or 3. Particles are generated only for those cells that have a *Mask* value > 0.

MaskLayer—is a specified layer used to define grid cells in a particle group when automatic particle generation is used and *GridCellRegionOption* = 3. When *MaskLayer* equals 0, particles are placed in the top-most active model layer at each areal grid cell location at the specified release time for each particle. If no active model layer exists at a specified particle release time, the particle is ignored.

FaceCount—is the number of cell faces that will have particles assigned to them by automatic particle generation when *PlacementOption* = 1. *FaceCount* has a value in the range 1 to 6.

IFace—is an index number used to associate a flow term with a grid cell face. *IFace* values between 1 and 6 correspond to the six grid cell faces as defined previously in this report. An *IFace* value of 0 is used in certain situations to denote a flow that is assumed to be distributed internally within a cell. An *IFace* value of -1 is used in certain situations to indicate that a flow should be distributed uniformly over boundary faces that are perpendicular to the lateral flow direction (faces 1 through 4).

ParticleLayerCount, *ParticleRowCount*, *ParticleColumnCount*—are the number of layer, row, and column subdivisions used to define arrays of particles within cells or on cell faces when automatic particle generation is used.

In the case of a three-dimensional array of particles, *ParticleLayerCount* is the number of vertical subdivisions, *ParticleRowCount* is the number of subdivisions in the y-direction, and *ParticleColumnCount* is the number of subdivisions in the x-direction.

In the case of a two-dimensional array of particles on a cell face, the following definitions apply: For faces 1 through 4, *ParticleRowCount* is the number of vertical subdivisions and *ParticleColumnCount* is the number subdivisions in the horizontal direction (the y-direction for faces 1 and 2; the x-direction for faces 3 and 4). For faces 5 and 6, *ParticleRowCount* is the number of subdivisions in the y-direction and *ParticleColumnCount* is the number of subdivisions in the x-direction.

Particle Output File Structure

The input data and results for MODPATH simulations are summarized in the Listing file. The MODPATH Listing file is similar in structure and appearance to the MODFLOW Listing file, and serves the same purpose.

MODPATH generates four types of particle output files: the endpoint, pathline, timeseries, and advective observations files. The endpoint file only contains information about the initial and final particle locations. The pathline file contains all the particle locations computed for every particle during the particle-tracking simulation. The timeseries file and the advective observations file contain particle location information for particles at user-specified time points. Particle output is controlled by specifying the value of *SimulationType* in the Simulation file:

SimulationType = 1:   An endpoint simulation. Endpoint simulations only generate an endpoint file.

SimulationType = 2:   A pathline simulation. Pathline simulations generate an endpoint file and a pathline file.

SimulationType = 3:   A timeseries simulation. Timeseries simulations generate an endpoint file and a timeseries file. Users also have the option of generating an advective observations file for timeseries simulations.

Particle output files are described in more detail in the next four sections.

In addition to the *Listing File* and the *Particle Output Files*, MODPATH also provides an option to generate a *Trace* file that provides detailed information about the particle-tracking history of a single particle specified by the user. Trace files are primarily intended to help analyze model results when problems are suspected. The option to generate a Trace file is specified in the MODPATH simulation file.

Endpoint File

The endpoint file is a text file that consists of a series of header lines followed by a sequence of one-line data records for each particle that was released during the simulation. All data are free-format. Multiple data items on a single line are separated by one or more spaces. The number of lines in the endpoint file header is equal to 5 plus the number of particle groups. An endpoint file is generated for every MODPATH simulation. When *SimulationType* equals 1, the endpoint file is the only particle output file generated. When *SimulationType* equals 2 or 3, a pathline or timeseries file is generated in addition to the endpoint file.

Header

Item 1   *Label   Version   Revision*

$\qquad$ *Label* =   MODPATH_ENDPOINT_FILE

$\qquad$ *Version* =   6

$\qquad$ *Revision* =   0

Item 2   *TrackingDirection   TotalCount   ReleaseCount   MaximumID   ReferenceTime*

$\qquad$ *TrackingDirection* :   1 = Forward, 2 = Backward

$\qquad$ *TotalCount* :   The total number of particles allocated for the simulation, including particles that may not actually be released as active during the simulation.

$\qquad$ *ReleaseCount*:   The number of particles that were actually released as active during the simulation. Only particles released as active are recorded in the endpoint file. Therefore, the number of particle records is always equal to the value of *ReleaseCount*.

$\qquad$ *MaximumID* :   The maximum particle ID value recorded in the file.

$\qquad$ *ReferenceTime* :   The reference time for the simulation.

Item 3   *Pending   Active   NormallyTerminated   ZoneTerminated   Unreleased   Stranded*
Values indicating the number of particles in each of the six status categories:

$\qquad$ *Pending* .   Status = 0. Particles that are scheduled to be released but have not yet been released. At the start of a simulation, all particles have a status of pending.

$\qquad$ *Active* :   Status = 1. Particles that are actively moving in the flow system and have not yet reached a termination location.

$\qquad$ *NormallyTerminated* :   Status = 2. Particles that have terminated at a boundary or internally at a cell with an internal source/sink.

$\qquad$ *ZoneTerminated* :   Status = 3. Particles that terminated at a cell with a specified zone number indicating automatic termination.

$\qquad$ *Unreleased* :   Status = 4. Particles that were not released and were tagged as permanently unreleased. The most common situation that results in unreleased particles is a dry or inactive cell condition at the scheduled release time.

$\qquad$ *Stranded* :   Status = 5. Particles that remain in cells after the cell goes dry. Stranded particles sometimes occur in transient simulations. Once a particle is stranded, it cannot be reactivated and is considered terminated.

Item 4   *GroupCount*
GroupCount is the number of particle groups.

Item 5   *GroupName*
Repeat data item 5 for each group.

Item 6   The end of the header is marked by the following line of text that begins in column 1:

END HEADER

Particle Endpoint Records

The endpoint file contains a particle record for each particle released during the simulation. A particle record consists of a single line of text that contains 30 space-delimited data items. Data items are defined as follows:

1. Particle ID	11. Initial Zone	21. Final Column
2. Particle Group	12. Initial Local X	22. Final Cell Face
3. Status	13. Initial Local Y	23. Final Zone
4. Initial Time	14. Initial Local Z	24. Final Local X
5. Final Time	15. Initial Global X	25. Final Local Y
6. Initial Grid	16. Initial Global Y	26. Final Local Z
7. Initial Layer	17. Initial Global Z	27. Final Global X
8. Initial Row	18. Final Grid	28. Final Global Y
9. Initial Column	19. Final Layer	29. Final Global Z
10. Initial Cell Face	20. Final Row	30. Label

The initial and final grid index values always equal 1, because the current version of MODPATH only supports a single finite-difference grid. Inclusion of the grid index variables allows the structure of the endpoint file to be compatible with future versions of MODPATH that may support multiple grids as implemented by the Local Grid Refinement (LGR) feature of MODFLOW.

Initial and final global elevation values are computed by MODPATH from the local vertical coordinate using the top and bottom elevations of the grid cell together with the assumption that the cell is a horizontal rectangular prism. That assumption is strictly valid only when the model layer containing the cell is horizontal. If the top or bottom of a model layer varies horizontally, it is more appropriate to use horizontal interpolation of the top and bottom elevation arrays to compute top and bottom elevations at the x-y location of the particle, and then use those values to compute the global vertical coordinate from the local vertical coordinate. Post-processing programs are required to generate the interpolated global vertical coordinates.

The data item, *Label*, is a user-defined text string with a maximum of 40 characters. Spaces are not permitted.

Pathline File

The pathline file is a text file that consists of a series of header lines followed by a sequence of one-line data records for each particle location computed during the MODPATH simulation. All data are free-format. Multiple data items on a single line are separated by one or more spaces. The file is produced for MODPATH simulations when the simulation type is set to pathline (*SimulationType* = 2).

Header

Item 1   *Label   Version   Revision*

          Label =   MODPATH_PATHLINE_FILE

          Version =   6

          Revision =   0

Item 2   *TrackingDirection   ReferenceTime*

          TrackingDirection :   1 = Forward, 2 = Backward

          ReferenceTime :   The reference time for the simulation.

Item 3   The end of the header is marked by the following line of text that begins in column 1:

          END HEADER

Particle Pathline Records

A particle location record consists of a single line of text containing 16 data items that are separated by one or more spaces. The data items are defined as follows:

1. Particle ID	11. Layer
2. Particle Group	12. Row
3. Time Point Index	13. Column
4. Cumulative Time Step	14. Grid
5. Tracking Time	15. Local X
6. Global X	16. Local Y
7. Global Y	17. Local Z
8. Global Z	18. Line Segment Index

The *Time Point Index* indicates whether a location record corresponds to a time point specified by the user. A time point value of 0 represents the starting time for the simulation. Subsequent time points are indicated by positive integers. Locations that do not represent a specified time point have a time point index equal to -1.

The *Cumulative Time Step* starts at 1 for the first time step of the MODFLOW simulation and increments sequentially through the last time step of the simulation.

Tracking Time is the value of time corresponding to the particle at the specified location.

MODPATH tracks particles within a time interval loop. The time intervals are defined by a combination of user-specified time points and the time breaks between MODFLOW time steps. Within a time interval, particles are tracked sequentially in a particle loop. For each particle, a record is written for the location at the start of the time interval, then additional records are written for the locations where the particle moves from one cell to another across a cell face. A final record is written for the location of the particle at the end of the time interval. The particle loop is repeated for each time interval. A MODPATH simulation that consists of multiple time intervals will result in a pathline file containing of multiple segments for a single particle. The multiple segments are indicated by the *Line Segment Index*.

Timeseries File

The timeseries file is a text file that consists of a series of header lines followed by a sequence of one-line data records of particle locations at the specified time points. All data are free-format. Multiple data items on a single line are separated by one or more spaces. The file is produced for MODPATH simulations when the simulation type is set to timeseries (*SimulationType* = 3).

Header

Item 1   *Label Version Revision*

> *Label* =   MODPATH_TIMESERIES_FILE
>
> *Version* =   6
>
> *Revision* =   0

Item 2   *TrackingDirection ReferenceTime*

> *TrackingDirection* :   1 = Forward, 2 = Backward
>
> *ReferenceTime* :   The reference time for the simulation.

Item 3   The end of the header is marked by the following line of text that begins in column 1:

END HEADER

Particle Timeseries Records

A particle location record consists of a single line of text containing 15 data items that are separated by one or more spaces. The data items are defined as follows:

1. Time Point Index
2. Cumulative Time Step
3. Tracking Time
4. Particle ID
5. Particle Group
6. Global X
7. Global Y
8. Global Z
11. Grid
12. Layer
13. Row
14. Column
15. Local X
16. Local Y
17. Local Z

Advective Observations File

The advective observations file is generated for timeseries simulations when the IADVOBS flag is set equal to 1 in the MODPATH simulation file. The advective observations file contains location information for each particle at the time points specified for the timeseries simulation. In contrast to the timeseries file, the advective observations file contains location data for all particles at all time points, even those that have already terminated internally or at grid boundaries. The location of a terminated particle for a time greater than its time of termination is computed by linear projection based on its velocity at the time of termination. Although the projected locations of terminated particles have no physical basis, they serve as placeholder values that permit parameter estimation applications to use advective observations to adjust travel time as a calibration target.

The advective observations file is a text file that contains a one-line data record for each particle at each time point. The data record is free format and contains the following nine data items:

1. Particle ID
2. Status
3. Global X
4. Global Y
5. Global Z
6. Tracking Time
7. VX
8. VY
9. VZ

VX, VY, and VZ are the velocity components for the particle at the time it was terminated. If the particle is still active (Status = 1), the values of the velocity components are set equal to 0.

Example Problem

The conceptual model of the flow system used in the example simulations is shown in Figure 8. The system consists of two aquifers separated by a 20-foot confining bed. The source of water for the system occurs as areal recharge to aquifer 1 at a constant, uniform rate of 0.005 foot per day (ft/d). Water discharges to a river on the right side of the system and a well located at the center of system that is screened in the bottom 100 feet of aquifer 2. The river stage is fixed at 320 feet throughout its entire length. The system is characterized by a period of steady-state flow with no pumping well followed by a period of transient flow in response to the addition of a discharge well pumping at a rate of 150,000 cubic feet per day (ft³/d). The pumping rate of the well is maintained at that level and the system eventually reaches a new steady state. The hydraulic properties and the porosity of the aquifers and confining bed are:

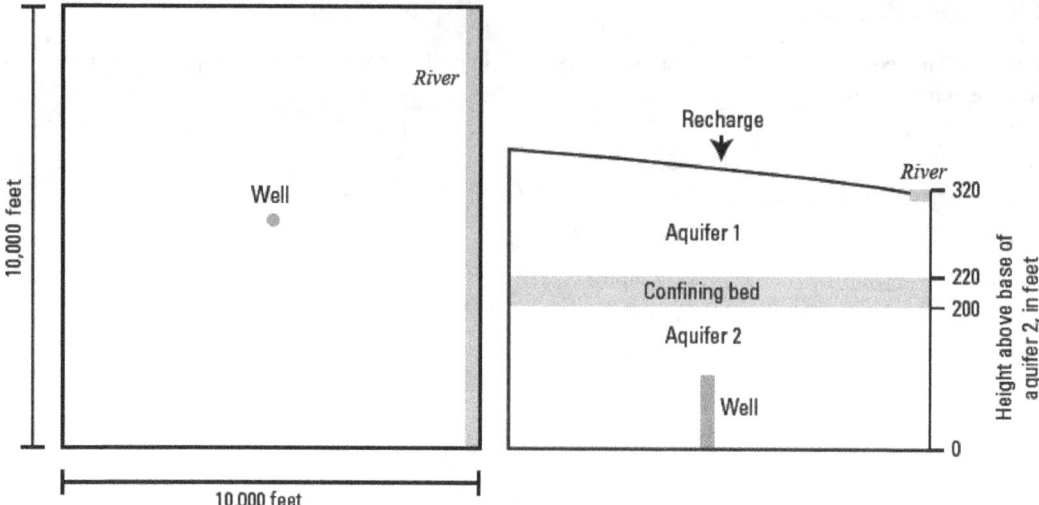

Figure 8.   Conceptual model for example problem.

Aquifer 1

- Horizontal hydraulic conductivity = 50 ft/d

- Vertical hydraulic conductivity = 10 ft/d

- Specific storage = 0.0001 foot $^{-1}$

- Specific yield = 0.2

- Porosity = 0.3

Aquifer 2

- Horizontal hydraulic conductivity = 200 ft/d

- Vertical hydraulic conductivity = 20 ft/d

- Specific storage = 0.0001 foot $^{-1}$

- Porosity = 0.3

Confining bed

- Horizontal hydraulic conductivity = 0.01 ft/d

- Vertical hydraulic conductivity = 0.01 ft/d

- Specific storage = 0.0001 foot $^{-1}$

- Porosity = 0.3

To represent this system with MODFLOW, the system was divided into five model layers. Layers 1 and 2 represent aquifer 1, layer 3 represents the confining bed, and layers 4 and 5 represent aquifer 2. The areal grid consists of 25 rows and 25 columns with a uniform cell size of 400 feet per side. The time discretization consists of three stress periods. Stress period 1 is a steady-state period 200,000 days long containing one time step. Stress period 2 is a transient period 10,000 days long containing 10 time steps. Stress period 3 is a steady-state period 200,000 days long containing one time step. The well is actively pumping in stress periods 2 and 3. Stress period 1 has no well discharge. The rate and distribution of recharge and the river stage are the same for all three stress periods. The MODFLOW input files are shown next:

```
MODFLOW Name File - example.nam
LIST              2 EXAMPLE.LST
DIS               3 EXAMPLE.DIS
BAS6              4 EXAMPLE.BA6
```

```
WEL               12 EXAMPLE.WEL
RIV               14 EXAMPLE.RIV
RCH               18 EXAMPLE.RCH
OC                22 EXAMPLE.OC
PCG               23 EXAMPLE.PCG
LPF               33 EXAMPLE.LPF
DATA(BINARY)      50 EXAMPLE.HED
DATA(BINARY)      53 EXAMPLE.BUD


Discretization (DIS) - example.dis
         5         25        25   3   4   1
 0 0 0 0 0
CONSTANT   4.000E+02
CONSTANT   4.000E+02
CONSTANT   4.000000E+02
CONSTANT   2.700000E+02
CONSTANT   2.200000E+02
CONSTANT   2.000000E+02
CONSTANT   1.000000E+02
CONSTANT   0.000000E+00
    2.000000E+05        1 1.000E+00  SS
    1.000000E+04       10 1.500E+00  TR
    2.000000E+05        1 1.000E+00  SS
Basic Package (BAS6) - example.ba6
FREE
CONSTANT          1
CONSTANT          1
CONSTANT          1
CONSTANT          1
CONSTANT          1
 1.000E+30
CONSTANT   3.200000E+02
CONSTANT   3.200000E+02
CONSTANT   3.200000E+02
CONSTANT   3.200000E+02
CONSTANT   3.200000E+02
Layer Property Flow Package (LPF) - example.lpf
        53  1.00E+20          0
 1 0 0 0 0
 0 0 0 0 0
  1.00E+00  1.00E+00  1.00E+00  1.00E+00  1.00E+00
 0 0 0 0 0
 0 0 0 0 0
CONSTANT   5.000000E+01
CONSTANT   1.000000E+01
CONSTANT   1.000000E-04
CONSTANT   2.000000E-01
CONSTANT   5.000000E+01
CONSTANT   1.000000E+01
CONSTANT   1.000000E-04
CONSTANT   1.000000E-02
CONSTANT   1.000000E-02
CONSTANT   1.000000E-04
CONSTANT   2.000000E+02
CONSTANT   2.000000E+01
CONSTANT   1.000000E-04
CONSTANT   2.000000E+02
CONSTANT   2.000000E+01
CONSTANT   1.000000E-04
Well Package (WEL) - example.wel
```

```
         1        53  AUXILIARY IFACE
         0         0
         1         0
         5        13         13 -1.500E+5          0
        -1         0
Recharge Package (RCH) - example.rch
         1        53
         0         0
CONSTANT   5.000000E-03
        -1         0
        -1         0


River Package (RIV) - example.riv
        25        53  AUXILIARY IFACE
        25         0
         1         1        25 3.200E+02 1.000E+05 3.150E+02          6
         1         2        25 3.200E+02 1.000E+05 3.150E+02          6
         1         3        25 3.200E+02 1.000E+05 3.150E+02          6
         1         4        25 3.200E+02 1.000E+05 3.150E+02          6
         1         5        25 3.200E+02 1.000E+05 3.150E+02          6
         1         6        25 3.200E+02 1.000E+05 3.150E+02          6
         1         7        25 3.200E+02 1.000E+05 3.150E+02          6
         1         8        25 3.200E+02 1.000E+05 3.150E+02          6
         1         9        25 3.200E+02 1.000E+05 3.150E+02          6
         1        10        25 3.200E+02 1.000E+05 3.150E+02          6
         1        11        25 3.200E+02 1.000E+05 3.150E+02          6
         1        12        25 3.200E+02 1.000E+05 3.150E+02          6
         1        13        25 3.200E+02 1.000E+05 3.150E+02          6
         1        14        25 3.200E+02 1.000E+05 3.150E+02          6
         1        15        25 3.200E+02 1.000E+05 3.150E+02          6
         1        16        25 3.200E+02 1.000E+05 3.150E+02          6
         1        17        25 3.200E+02 1.000E+05 3.150E+02          6
         1        18        25 3.200E+02 1.000E+05 3.150E+02          6
         1        19        25 3.200E+02 1.000E+05 3.150E+02          6
         1        20        25 3.200E+02 1.000E+05 3.150E+02          6
         1        21        25 3.200E+02 1.000E+05 3.150E+02          6
         1        22        25 3.200E+02 1.000E+05 3.150E+02          6
         1        23        25 3.200E+02 1.000E+05 3.150E+02          6
         1        24        25 3.200E+02 1.000E+05 3.150E+02          6
         1        25        25 3.200E+02 1.000E+05 3.150E+02          6
        -1         0
        -1         0
Output Control (OC) - example.oc (partial listing)
HEAD PRINT FORMAT    0
DRAWDOWN PRINT FORMAT   0
HEAD SAVE UNIT  50
COMPACT BUDGET AUXILIARY
PERIOD   1 STEP    1
     PRINT BUDGET
     SAVE BUDGET
     PRINT HEAD
     SAVE HEAD
PERIOD   2 STEP    1
     SAVE BUDGET
     PRINT HEAD
     SAVE HEAD
PERIOD   2 STEP    2
     SAVE BUDGET
     PRINT HEAD
     SAVE HEAD
```

```
PERIOD   2 STEP    3
    SAVE BUDGET
    PRINT HEAD
    SAVE HEAD

...
Solver Package (PCG) - example.pcg
        15        20         1
   1.00E-04  1.00E+03  1.00E+00         2         1         0 1.000E+00
```

The output control file (EXAMPLE.OC) is set to save cell-by-cell flow output in compact budget form with auxiliary data included. Flow output from the LPF, RCH, RIV, and WEL packages is saved into a single output file (EXAMPLE.BUD). The river package and well package files include IFACE values as an auxiliary data item. IFACE values appear as the last data item on the line for the well and each river reach. In the case of the well, the IFACE value equals 0, which indicates that MODPATH should treat the well as a distributed internal sink. The IFACE value for each river reach is set equal to 6 to indicate that the flow output to the river reach should be assigned to the top face of the cell and used to compute a vertical velocity component at the cell face.

Figure 9 shows the head distribution in layers 1 and 5 for the steady-state flow period in stress period 3, time step 1. The head distribution in layer 1 is relatively unaffected by the well in layer 5 due to the isolating effect of the confining bed represented by layer 3. The groundwater flow system just described forms the basis for five MODPATH particle-tracking simulations that illustrate the major types of analyses that can be performed with MODPATH.

Once the MODFLOW simulation has been completed, the next step is to prepare the MODPATH data files. In addition to the MODFLOW discretization, budget, and head files, each MODPATH simulation requires these additional input files:

- MODPATH name file
- MODPATH basic data file
- MODPATH simulation file

All of the MODPATH simulations for this example problem use the following MODPATH name file and MODPATH basic data file:

```
MODPATH Name File - example.mpnam
MPBAS              70 EXAMPLE.MP6
DIS                3 EXAMPLE.DIS
HEAD               1 EXAMPLE.HED
BUDGET             2 EXAMPLE.BUD
MODPATH Basic Data (MPBAS) - example.mp6
  1.000E+30  1.00E+20
          1
RECHARGE
          6
  1 0 0 0 0
CONSTANT          1
CONSTANT          1
CONSTANT          1
CONSTANT          1
CONSTANT          1
CONSTANT  3.000000E-01
CONSTANT  3.000000E-01
CONSTANT  3.000000E-01
CONSTANT  3.000000E-01
CONSTANT  3.000000E-01
```

The MODPATH name file is similar to the MODFLOW name file, and it provides file connection information for many of the primary data files needed by MODPATH. The MODPATH basic data file consolidates information from the MODFLOW basic and flow-package data files that is needed by MODPATH, as well as porosity data, which is not required by MODFLOW. The MODPATH basic data file usually will not change unless the structure of the MODFLOW simulation changes.

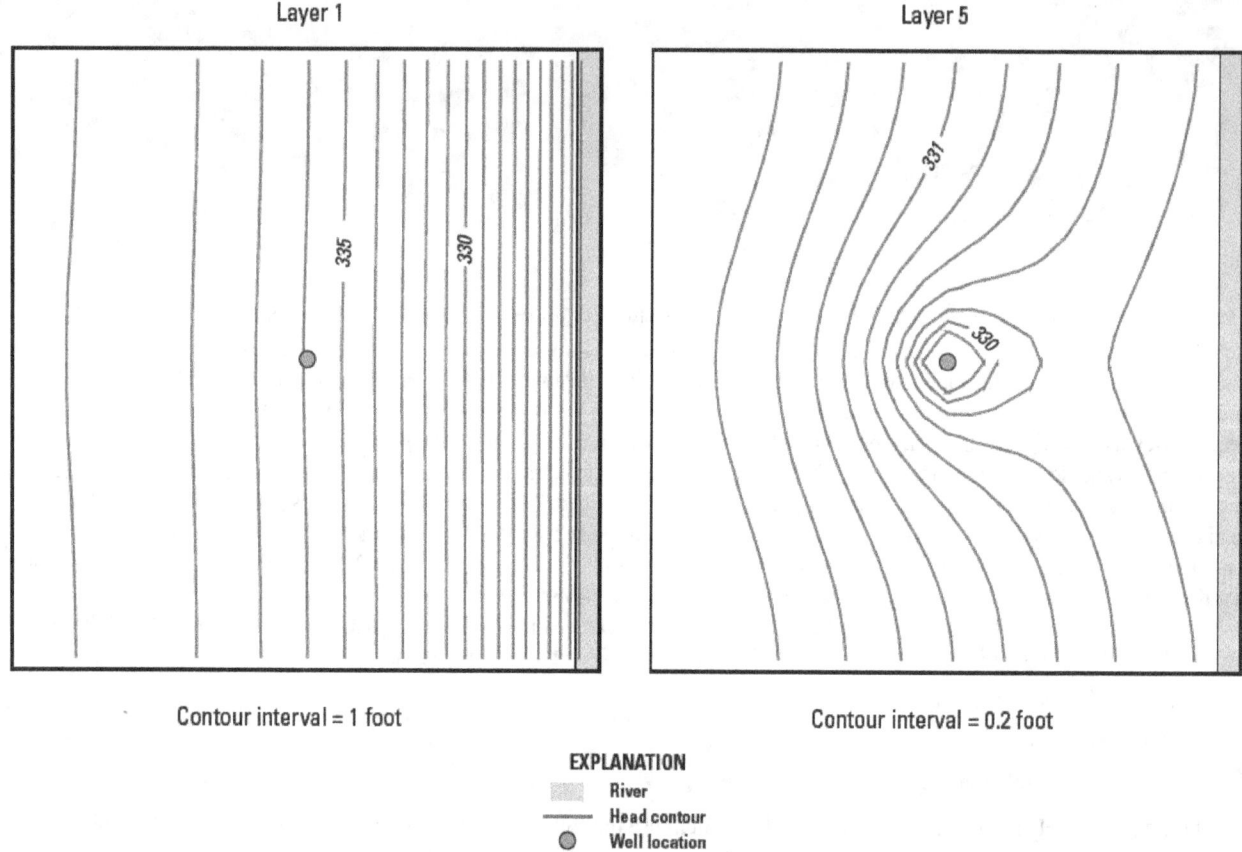

Layer 1

Layer 5

Contour interval = 1 foot

Contour interval = 0.2 foot

EXPLANATION

River

Head contour

Well location

Figure 9.   Head in layers 1 and 5 for steady-state flow conditions in stress period 3, time step 1.

For any given MODFLOW simulation, it is common to generate several MODPATH simulations designed to examine the advective characteristics of the flow system from a variety of perspectives. The data that customize MODPATH for each specific particle-tracking simulation are found in the MODPATH simulation file.

Simulation 1—Forward Endpoint Simulation

This simulation is designed to determine the source area for water that discharges to the pumping well in layer 5, row 13, column 13 under the steady-state flow conditions in stress period 3. A 2×2 array of particles is placed on the top face of each cell in layer 1 to cover the entire recharge area. A total of 2,500 particles are released at the beginning of stress period 3, time step 1 and tracked forward to their discharge locations. The simulation is run in endpoint mode so that only the initial and final locations of the particles are recorded in the output. The source area for the well can be determined by mapping the starting locations of all the particles that terminate in the cell that contains the well. The particle starting locations are generated using the automatic particle generation option for a rectangular block of cells corresponding to layer 1. The option to extend the final steady-state time step is selected to assure that all of the particles will be tracked to their termination points. The simulation file is shown below.

Simulation File 1 – example-1 mpsim
```
EXAMPLE.mpnam
EXAMPLE-1.mplist
  1  1  1  1  2  2  1  1  2  2  1  1
EXAMPLE-1.endpoint
```

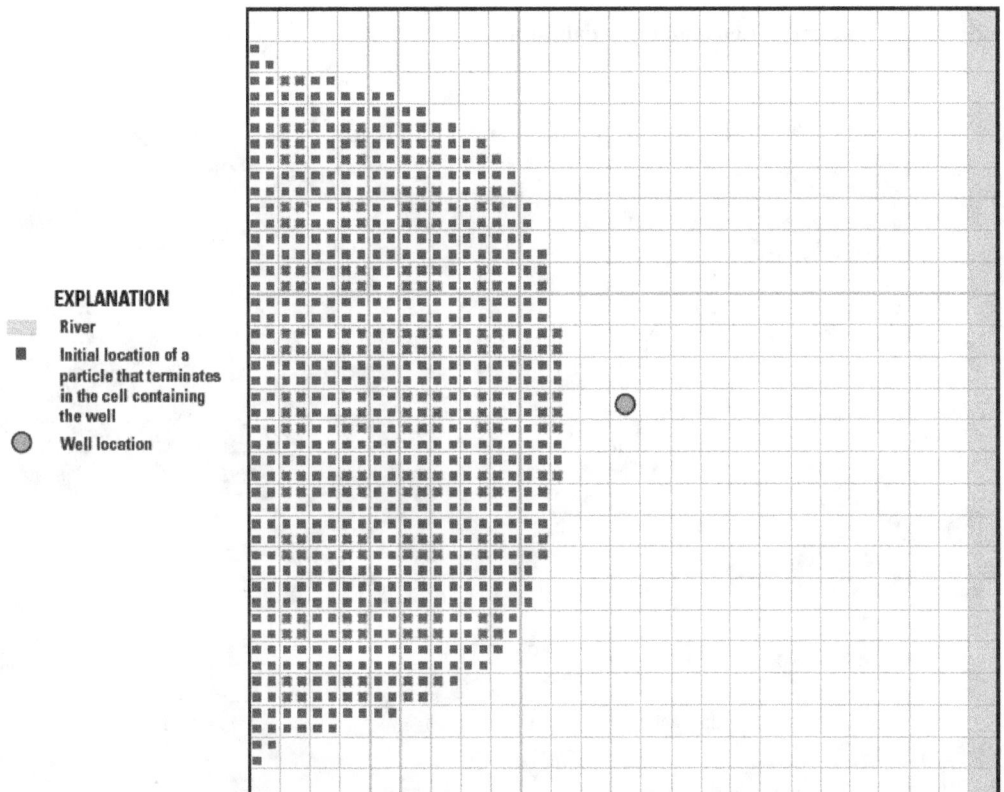

Figure 10. Simulation 1—Forward tracking endpoint simulation.

EXPLANATION
River
Initial location of a
particle that terminates
in the cell containing
the well
Well location

```
    3    1    0.00000
    1
PG00
    1    1    1    0.000000E+00    1    1
    1    1    1    1    25    25
    1
    6    2    2
    0
CONSTANT          1
CONSTANT          1
CONSTANT          1
CONSTANT          1
CONSTANT          2
```

When the MODPATH program is executed, it will prompt the user to enter the name of the simulation file. As an alternative, the simulation filename can be entered as a command-line argument. The specific procedure for running MODPATH depends on the computer operating system and how the software is installed. Consult the readme file that accompanies the software for details.

Figure 10 shows the starting locations of all the particles that terminated in the cell containing the well. Starting locations for particles that terminate in the river are not displayed.

The output data in the endpoint file (example-1.endpoint) was used to generate figure 10. The endpoint file and the other MODPATH particle coordinate output files are designed to provide data to other post-processing software, such as graphics display or statistical analysis programs. An excerpt from the endpoint file is shown below to provide an example of its structure. The file consists of a header section, followed by a data section composed of one line of data for each particle.

MODPATH endpoint file – example-1.endpoint (partial listing)

```
MODPATH_ENDPOINT_FILE 6 0
 1      2500      2500      2500 0.2100000E+06
        0         0      2500         0         0         0
        1
PG00
END HEADER
            1   1  2  0.0000000E+00  0.1894045E+06  1   1   1   1 6        1  0.2500000E+00  0.2500000E+00  ...
            2   1  2  0.0000000E+00  0.1469220E+06  1   1   1   1 6        1  0.7500000E+00  0.2500000E+00  ...
            3   1  2  0.0000000E+00  0.1853681E+06  1   1   1   1 6        1  0.2500000E+00  0.7500000E+00  ...
            4   1  2  0.0000000E+00  0.1449049E+06  1   1   1   1 6        1  0.7500000E+00  0.7500000E+00  ...
            5   1  2  0.0000000E+00  0.1275559E+06  1   1   1   2 6        1  0.2500000E+00  0.2500000E+00  ...
            6   1  2  0.0000000E+00  0.1148192E+06  1   1   1   2 6        1  0.7500000E+00  0.2500000E+00  ...
            7   1  2  0.0000000E+00  0.1260649E+06  1   1   1   2 6        1  0.2500000E+00  0.7500000E+00  ...
            8   1  2  0.0000000E+00  0.1135930E+06  1   1   1   2 6        1  0.7500000E+00  0.7500000E+00  ...
            9   1  2  0.0000000E+00  0.1052610E+06  1   1   1   3 6        1  0.2500000E+00  0.2500000E+00  ...
           10   1  2  0.0000000E+00  0.9753173E+05  1   1   1   3 6        1  0.7500000E+00  0.2500000E+00  ...

...

         2495   1  2  0.0000000E+00  0.3203577E+03  1   1  25  24 6        1  0.2500000E+00  0.7500000E+00  ...
         2496   1  2  0.0000000E+00  0.1033661E+03  1   1  25  24 6        1  0.7500000E+00  0.7500000E+00  ...
         2497   1  2  0.0000000E+00  0.0000000E+00  1   1  25  25 6        1  0.2500000E+00  0.2500000E+00  ...
         2498   1  2  0.0000000E+00  0.0000000E+00  1   1  25  25 6        1  0.7500000E+00  0.2500000E+00  ...
         2499   1  2  0.0000000E+00  0.0000000E+00  1   1  25  25 6        1  0.2500000E+00  0.7500000E+00  ...
         2500   1  2  0.0000000E+00  0.0000000E+00  1   1  25  25 6        1  0.7500000E+00  0.7500000E+00  ...
```

MODPATH also generates a listing file that contains summary information for the simulation. An excerpt of the listing file for this simulation (example-1 mplist) is shown below.

```
MODPATH lising file - example-1.mplist (partial listing)
...

_____

Processing Time Step     1 Period    3.  Time =   4.1                 (Cumulative Step =    12)
_____

PROCESSING HEAD AND BUDGET DATA FOR GRID    1
READ HEAD FOR PERIOD     3 STEP     1 PERTIM =   2.000000E+05  TOTIM =   4.100000E+05
UPDATE IBOUND FOR NO-FLOW AND DRY CELLS.
PROCESS FLOW PACKAGE BUDGET DATA FOR STRESS PERIOD    3 , TIME STEP     1 ...
Reading     CONSTANT HEAD for Period    3    Time Step    1
Reading FLOW RIGHT FACE  for Period    3    Time Step    1
Reading FLOW FRONT FACE  for Period    3    Time Step    1
Reading FLOW LOWER FACE  for Period    3    Time Step    1
Reading            WELLS for Period    3    Time Step    1
Reading    RIVER LEAKAGE for Period    3    Time Step    1
Reading         RECHARGE for Period    3    Time Step    1
         RECHARGE WILL BE ASSIGNED TO THE TOP FACE.


CHECKING BUDGET DATA FOR GRID    1
   NUMBER OF VARIABLE HEAD CELLS =      3125
          77 CELLS HAD BALANCE ERRORS BETWEEN 0.01 AND 0.1 PERCENT
           0 CELLS HAD BALANCE ERRORS BETWEEN 0.1 AND 1.0 PERCENT
           0 CELLS HAD BALANCE ERRORS BETWEEN 1.0 AND 10.0 PERCENT
           0 CELLS HAD BALANCE ERRORS BETWEEN 10.0 AND 50.0 PERCENT
           0 CELLS HAD BALANCE ERRORS GREATER THAN 50.0 PERCENT

   A MAXIMUM ERROR OF    0.039794 PERCENT OCCURRED IN LAYER    3  ROW   25  COLUMN   19
```

```
INDIVIDUAL CELL BUDGETS:
  CELL WITH MAXIMUM VOLUMETRIC BALANCE ERROR:
  GRID  1   LAYER    3   ROW   25   COLUMN   19
    FACE FLOWS:
          QX1 =   0.1193805E+00(IN)      QX2 =   0.1337330E+00 (OUT)
          QY1 =   0.0000000E+00          QY2 =   0.7145265E-03 (OUT)
          QZ1 =  -0.4012540E+02(OUT)     QZ2 =  -0.4012445E+02 (IN)
    CELL FACE FLOWS:  FLOW IN =   0.4024384E+02     FLOW OUT =   0.4025985E+02   NET INFLOW (IN - OUT) =  -0.1601791E-01
    FLOW TO SINKS =   0.0000000E+00   FLOW FROM SOURCES =   0.0000000E+00   FLOW FROM STORAGE =   0.0000000E+00
    VOLUMETRIC RESIDUAL =  -0.1601791E-01   VOLUMETRIC BALANCE(%) =    0.0397942


Particle summary:
─────────

       0 particles remain active
    1748 particles terminated in zone    1
     752 particles terminated in zone    2
       0 particles were stranded

End of MODPATH simulation. Normal termination.
```

A total of 752 particles terminate in the cell with the well (denoted as zone 2). If it is assumed that each particle represents one quarter of the recharge to each starting cell (200 cubic feet per day [ft^3/d]), the total recharge associated with the particles terminating in the well cell can be computed by multiplying 752 by 200 to obtain 150,400 ft^3/day. That value is within approximately 0.25 percent of the pumping rate specified in the MODFLOW simulation (150,000 ft^3/d). The presence of weak sinks and sources in actual problems makes it much more difficult to compute quantitative water balances based on particle-tracking output. Nevertheless, the fact that quantitative balances can be computed for simple problems such as this one provides a measure of confirmation that the particle-tracking method implemented by MODPATH is consistent with the distribution of groundwater flow produced by MODFLOW.

Simulation 2—Backward Endpoint Simulation

This simulation uses backward tracking to determine the recharge area for cell (5, 13, 13) that contains the pumping well for the steady-state flow conditions of stress period 3. A 10×10 array of particles is placed on each of the inflow faces of the cell (faces 1, 2, 3, 4, and 6). No particles are places on face 5 because it represents the bottom no-flow boundary of the aquifer. In contrast to simulation 1, the particles are released at the end of stress period 3, time step 1, so that they will have time to reach their termination point within the steady-state flow period of stress period 3. The simulation file is shown below.

Simulation File 2 – example-2.mpsim

```
EXAMPLE.mpnam
EXAMPLE-2.mplist
    1    2    1    1    2    2    1    1    2    2    1    1
EXAMPLE-2.endpoint
    3    1   1.00000
    1
PG00
    1    1    1   0.000000E+00    1    1
    5   13   13    5   13   13
    5
    1   10   10
    2   10   10
    3   10   10
    4   10   10
    6   10   10
    0
CONSTANT        1
CONSTANT        1
CONSTANT        1
CONSTANT        1
CONSTANT        2
```

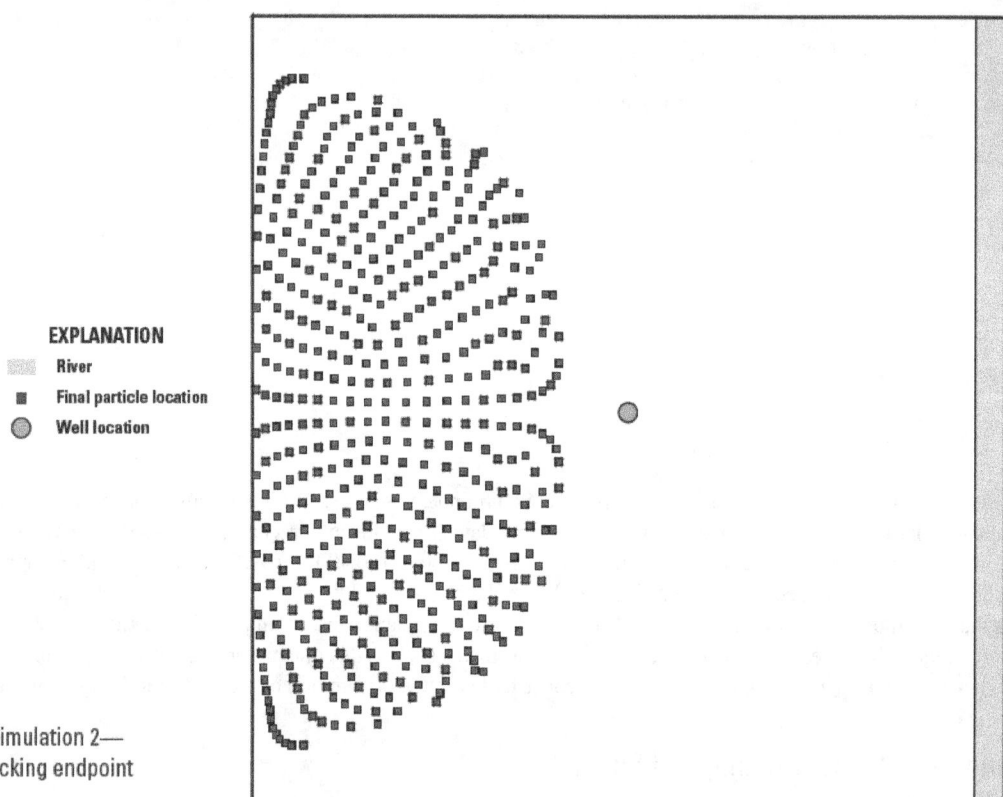

EXPLANATION

River

Final particle location

Well location

Figure 11.   Simulation 2—
backward tracking endpoint
simulation.

Figure 11 shows the final location of the particles, which all terminate at the recharge boundary at the top of layer 1. Comparison of figures 10 and 11 illustrate that the forward and backward tracking simulations map out the same recharge area.

Simulation 3—Forward Pathline Simulation

When MODPATH is run in endpoint mode only the initial and final locations of the particles are saved in an endpoint file. To draw pathlines, it is necessary to run MODPATH in pathline mode so that intermediate particle locations needed to define paths also are saved as output. The difference between endpoint and pathline simulations is only the amount of output that is saved. The internal particle-tracking computations are identical. In this simulation, a line of particles is placed at the top cell face of layer 1 in column 5 for rows 1 through 25. One particle is placed in each cell for a total of 25 particles. As in simulation 1, particles are released at the start of stress period 3, time step 1 and tracked to their termination point. The option to extend the final steady-state time step also is selected to assure that all particles are tracked to their termination points. The simulation file is shown below.

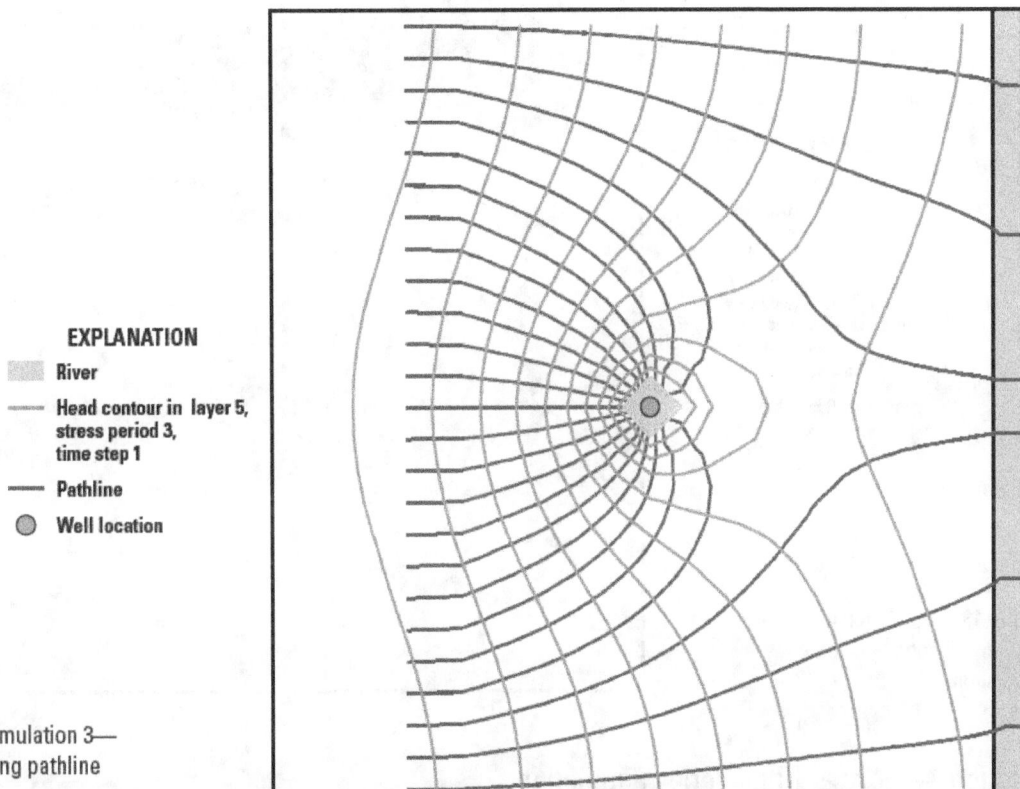

Figure 12. Simulation 3—
forward tracking pathline
simulation.

EXPLANATION

■ River

— Head contour in layer 5,
    stress period 3,
    time step 1

— Pathline

● Well location

Simulation File 3 – example-3.mpsim

```
EXAMPLE.mpnam
EXAMPLE-3.mplist
    2    1    1    1    2    2    1    1    2    2    1    1
EXAMPLE-3.endpoint
EXAMPLE-3.pathline
    3    1    0.00000
    1    #NPGROUPS
PG00
    1    1    1    0.000000E+00    1    1
    1    1    5    1    25    5
    1
    6    1    1
    0
CONSTANT        1
CONSTANT        1
CONSTANT        1
CONSTANT        1
CONSTANT        2
```

For pathline simulations, MODPATH generates both an endpoint and a pathline coordinate file. Figure 12 shows the results of
this simulation.

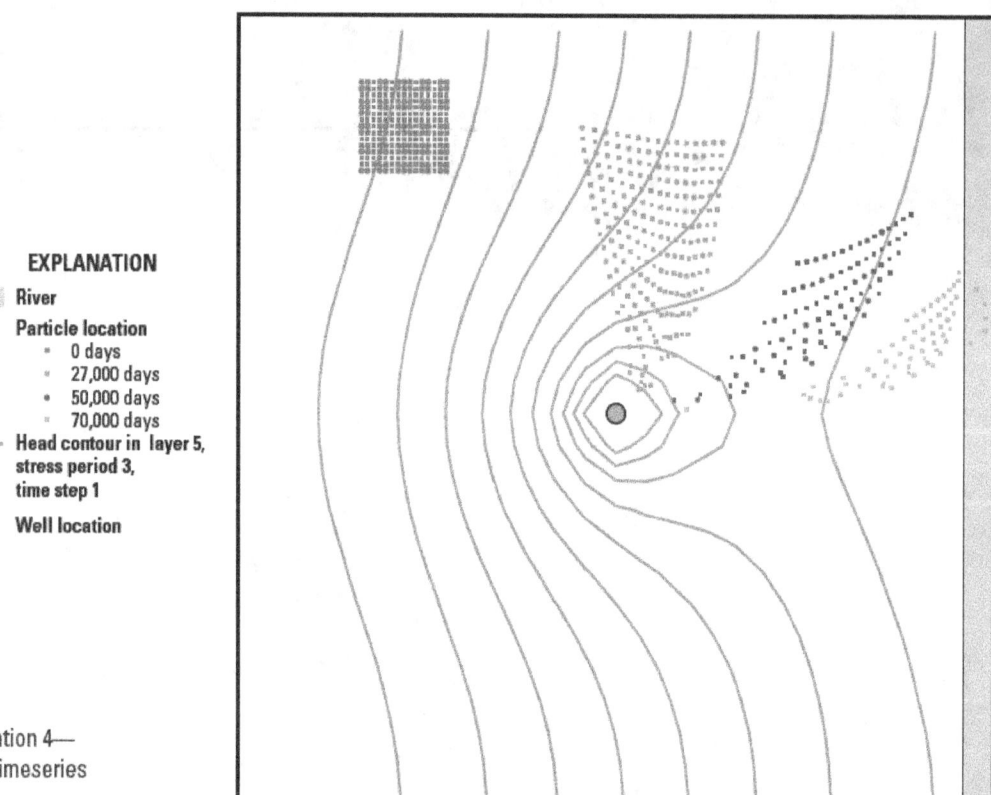

EXPLANATION

River

Particle location
   0 days
   27,000 days
   50,000 days
   70,000 days
—— Head contour in layer 5, stress period 3, time step 1

Well location

Figure 13.   Smulation 4—forward tracking timeseries simulation.

Simulation 4—Forward Timeseries Simulation

The final type of MODPATH simulation is a timeseries simulation. A timeseries simulation is designed to record particle locations at specific points in time so that particle coordinate output is available in the form of a series of snapshots in time. Timeseries simulations are useful for looking at groups of particles as they move through the flow system. This simulation illustrates a timeseries simulation in the forward direction. A 5×5 array of particles is placed on the top face of each cell in layer 1 for rows 3 through 5 and columns 5 through 7, yielding a total of 225 particles. The particles are released at the beginning of stress period 3. The locations of the particles are recorded in the timeseries file at constant 1,000-day time intervals. A total of 100 time intervals are recorded. The simulation file is shown below and results for four time points are presented in figure 13.

Simulation File 4 – example-4 mpsim

```
EXAMPLE.mpnam
EXAMPLE-4.mplist
    3    1    1    1    2    2    1    2    2    2    1    1
EXAMPLE-4.endpoint
EXAMPLE-4.timeseries
    3    1    0.00000
    1   #NPGROUPS
PG00
    1    1    1    0.000000E+00    1    1
    1    3    5    1    5    7
    1   #Nfaces
    6    5    5
  100   #NTPTS
    1.000000E+03
    0
CONSTANT          1
CONSTANT          1
CONSTANT          1
CONSTANT          1
CONSTANT          2
```

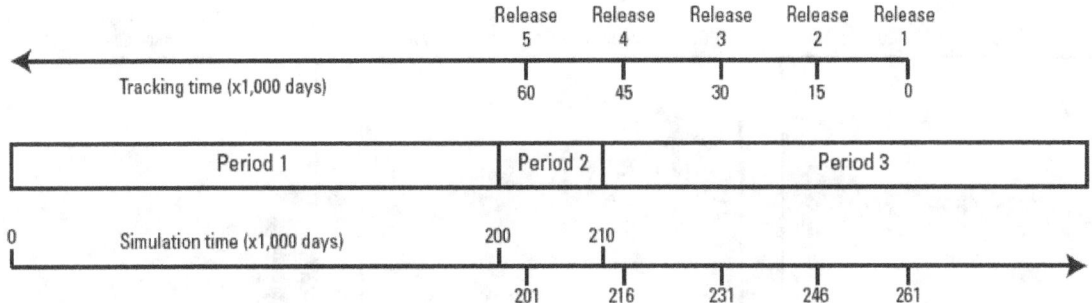

Figure 14. Multiple release times for simulation 5.

Simulation 5—Backward Multiple Release Endpoint Simulation

This simulation illustrates the use of multiple release times to define the change in the recharge area for the well cell over time in response to the onset of pumping at the start of stress period 2. A backward endpoint simulation is performed in which a group of particles is placed on the inflow faces of cell (5, 13, 13) using the same particle distribution as described for simulation 2. The reference simulation time is set equal to 261,000 days. Particles are released at tracking times of 0, 15,000, 30,000, 45,000, and 60,000 days. Figure 14 shows the relationship of the particle release time to the MODFLOW simulation time. When particles within each release batch are backtracked to their points of recharge at the water table, the result maps the recharge area for the well cell for the water discharging to the well cell at the time of the release.

The simulation file is show below.

Simulation File 5 – example-5.mpsim

```
EXAMPLE.mpnam
EXAMPLE-5.mplist
    1    2    1    1    1    2    1    1    2    2    1    1
EXAMPLE-5.endpoint
    2.610000E+05
    1
PG00
    1    1    1    0.000000E+00    2    1
    6.000000E+04    4
    5   13   13    5   13   13
    5
    1   10   10
    2   10   10
    3   10   10
    4   10   10
    6   10   10
    0
CONSTANT        1
CONSTANT        1
CONSTANT        1
CONSTANT        1
CONSTANT        2
```

Figure 15 shows the change in recharge area as a function of time for release times of 0, 45,000, and 60,000 days. The recharge area delineated in figure 15*A* is nearly identical to that shown for simulation 2 in figure 11. As the release times move closer to the start of pumping at the beginning of stress period 2 (fig. 15*B–C*), the recharge area decreases in size, a reflection of the fact that the particles released at those times spent progressively longer periods of their flow history under pre-pumping conditions.

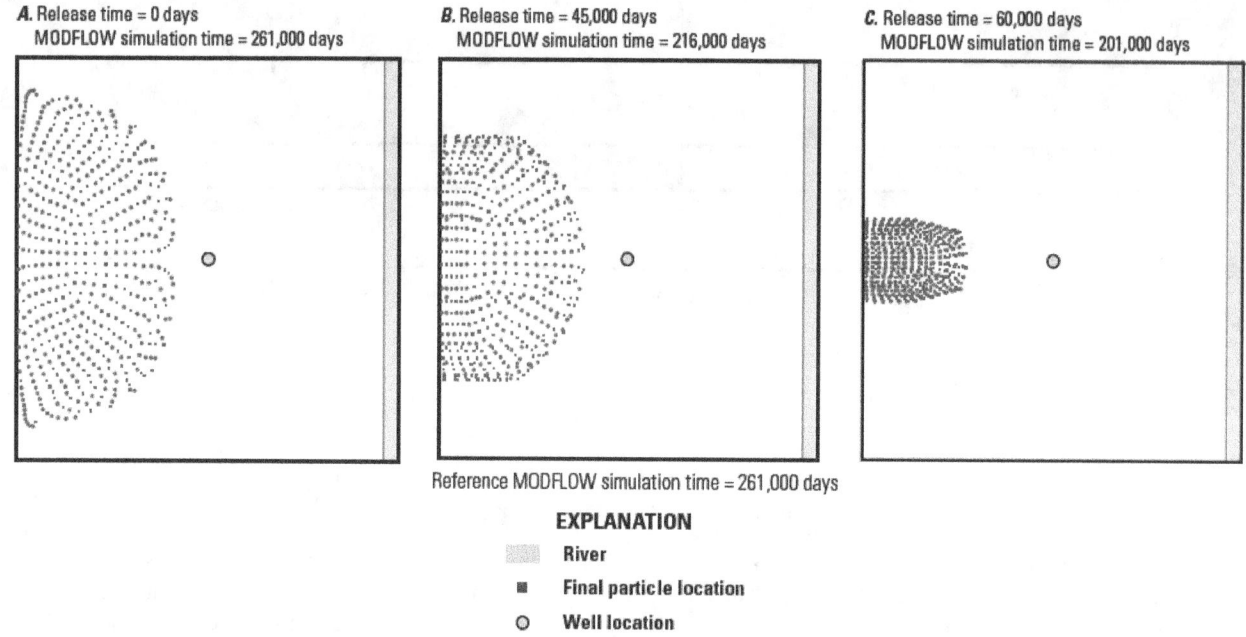

A. Release time = 0 days
   MODFLOW simulation time = 261,000 days

B. Release time = 45,000 days
   MODFLOW simulation time = 216,000 days

C. Release time = 60,000 days
   MODFLOW simulation time = 201,000 days

Reference MODFLOW simulation time = 261,000 days

EXPLANATION

River

■ Final particle location

○ Well location

Figure 15.   Simulation 5—backward tracking endpoint with multiple release times.

References

Domenico, P.A., and Schwartz, F.W., 1990, Physical and chemical hydrogeology: John Wiley and Sons, 824 p.

Harbaugh, A.W., 2005, MODFLOW-2005, the U.S. Geological Survey modular ground-water model—The Ground-Water Flow Process: U.S. Geological Survey Techniques and Methods 6–A16, variously paginated.

International Standards Organization, ISO/TEC 1539-1,1997, Information technology—Programming languages—Fortran— Part 1: Base language: Geneva (Fortran 95).

International Standards Organization, ISO/TEC 1539-1, 2004, Information technology—Programming languages—Fortran— Part 1: Base language: Geneva, 2004 (Fortran 2003).

Mehl, S.W., and Hill, M.C., 2005, MODFLOW-2005, The U.S. Geological Survey modular ground-water model—Documentation of shared node local grid refinement (LGR) and the boundary flow and head (BFH) package: U.S. Geological Survey Techniques and Methods 6–A12, 68 p.

Pollock, D.W., 1989, Documentation of a computer program to compute and display pathlines using results from the U.S. Geological Survey modular three-dimensional finite-difference ground-water flow model: U.S. Geological Survey Open-File Report 89–381, 188 p.

Pollock, D.W., 1994, User's guide for MODPATH/MODPATH-PLOT, version 3: A particle-tracking post-processing package for MODFLOW, the U.S. Geological Survey finite-difference ground-water flow model: U.S. Geological Survey Open-File Report 94–464, 249 p.

Toth, J., 1963, A theoretical analysis of groundwater flow in small drainage basins: Journal of Geophysical Research, v. 68, no. 16, p. 4795-4812.

Appendix 1.   MODPATH Output Examiner

Particle coordinate output generated by MODPATH is difficult to evaluate without the assistance of post-processing tools to help visualize and examine the output. *MODPATH Output Examiner* is a post-processing program that allows output from MODPATH version 6 to be visualized in a variety of ways. The program is a Microsoft Windows application that will run on computers running the Microsoft Windows operating system with the Microsoft .NET framework version 3.5 or later.

Installation

The MODPATH Output Examiner application consists of a number of executable component files that are grouped together in a single folder. The installation process simply involves copying that folder to the desired location on the user's computer. Although the folder can be renamed, all of the files contained in the folder must be kept together and their original names should not be changed. The required version of the Microsoft .NET Framework (version 3.5 or later) will already be installed on most computers running the Microsoft Windows operating system. If necessary, however, it can be downloaded from the Microsoft Web site and installed. Detailed instructions for setting up MODPATH Output Examiner are provided with the software distribution.

Starting the Application and Loading Data

To run MODPATH Output Examiner, double-click the executable application file *ModpathOutputExaminer.exe*. For convenience, it may be useful to create a desktop shortcut for the executable file that can be used to start the application. When the application starts, the main application screen will appear with no data loaded (fig. 1–1).

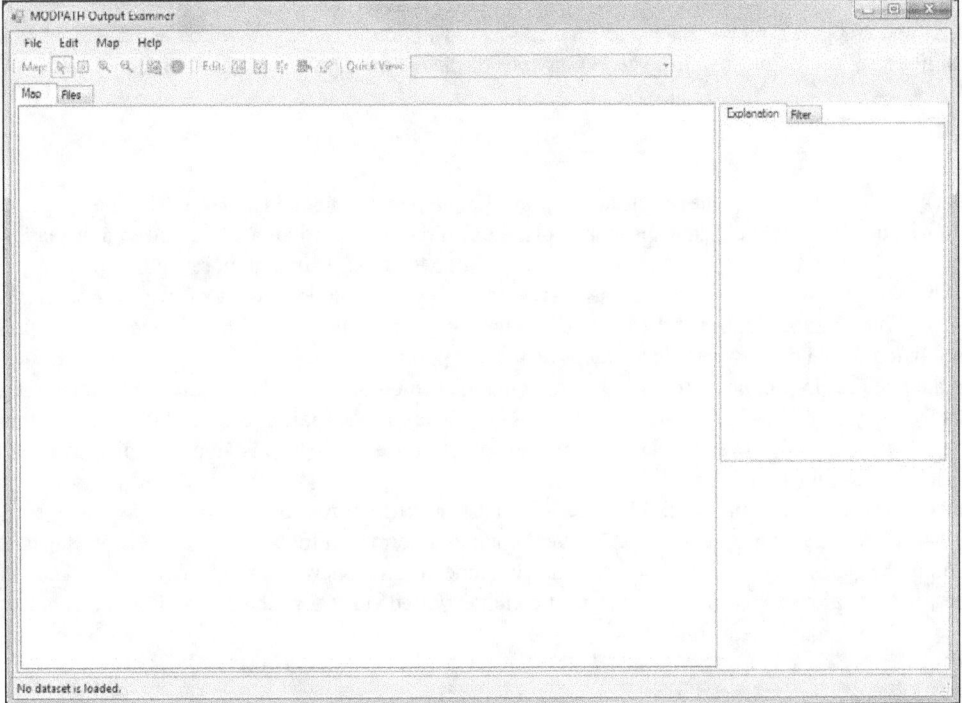

Figure 1–1.   Application screen with no dataset loaded.

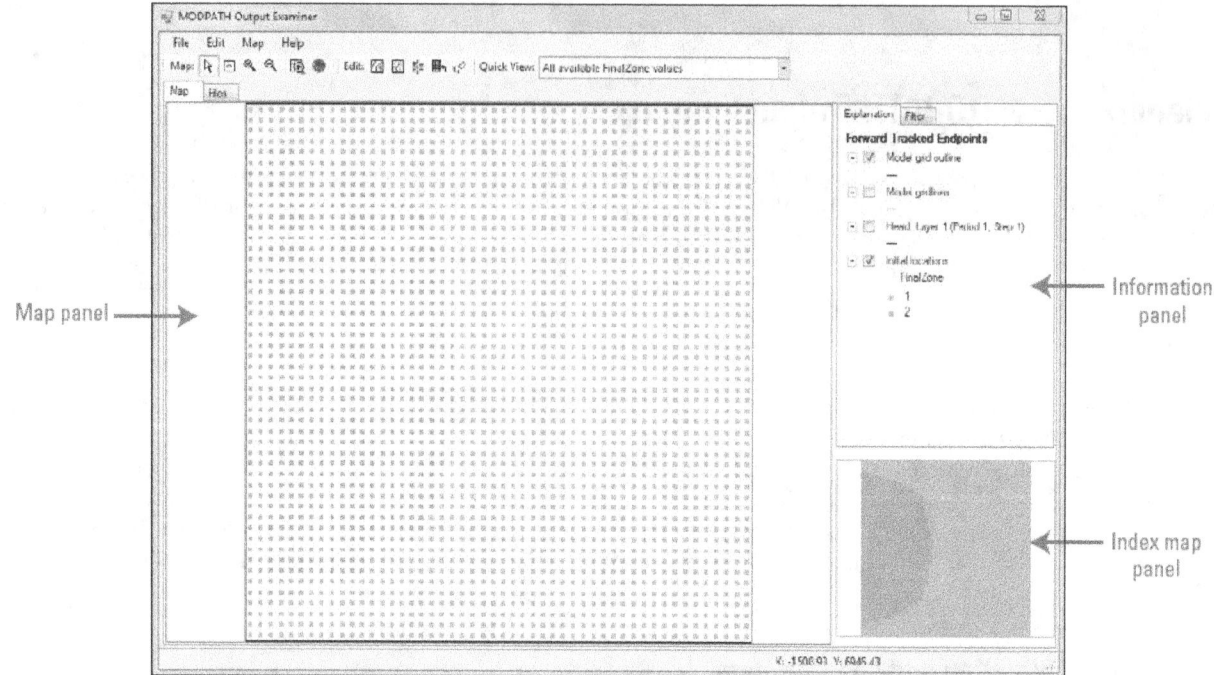

Figure 1–2. Application screen with dataset *example-1.mpsim* loaded.

MODPATH Output Examiner is designed to display results from specified MODPATH simulations. To load a simulation dataset, go to the *File* menu and select the option *Open Simulation*. Browse the file-open dialog and select a MODPATH simulation file. If a desktop shortcut to the application exists, it is also possible to load a simulation file dataset by dragging the simulation file from Windows Explorer and dropping it on the desktop shortcut. As an illustration, selecting the simulation file for example problem 1 (example-1.mpsim) causes the output from that simulation to be displayed in map view as shown in figure 1–2.

Map View

When a dataset is initially loaded, the results are displayed in map view, which includes a main map panel, an index map panel, and an information panel that contains the map explanation and other information about the data display.

The type of data displayed in map view is based on the type of MODPATH simulation (endpoint, pathline, or timeseries). In the case shown above, the simulation was a forward-tracking endpoint simulation, so the display shows initial locations of particles color-coded by the zone number of the final cell. In the case of a pathline simulation, the display would show pathlines. A timeseries simulation would display particle locations at specific points in time.

The appearance of the plot can be customized by changing the symbology (see Map Symbology Dialog section), and by filtering the data by applying constraints to particle attributes (see Query Filter Dialog section). In addition to the customizations that can be achieved through the edit dialogs, views can be customized by selecting an option directly from the drop-down Quick View list in the tool bar (fig. 1–3).

The display options provided in the Quick View drop-down list are customized for the type of data displayed. The items shown in figure 1–3 are for an endpoint simulation. Different options are provided for pathline and timeseries data. As an example, for timeseries output the Quick View provides tools for selecting time points to view and for displaying time animations. The goal of the Quick View list is to provide easy access to the most commonly used views for each of the MODPATH output types.

Files View

In addition to the map view, the primary MODPATH input files and the MODPATH listing output file can be viewed by clicking the files tab to switch to files view (fig. 1–4). In files view, select a file in the file list panel to view its contents below in the file contents panel.

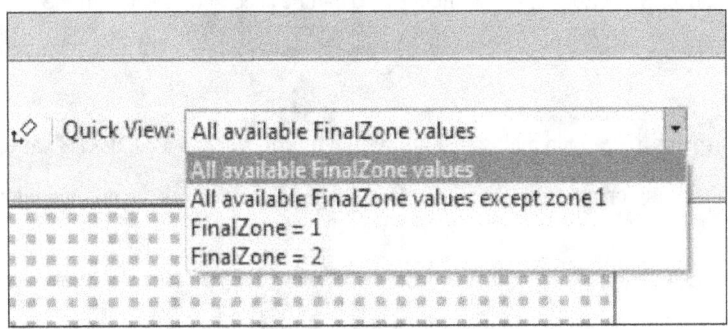

Figure 1–3.   Quick View drop-down list for example-1.mpsim.

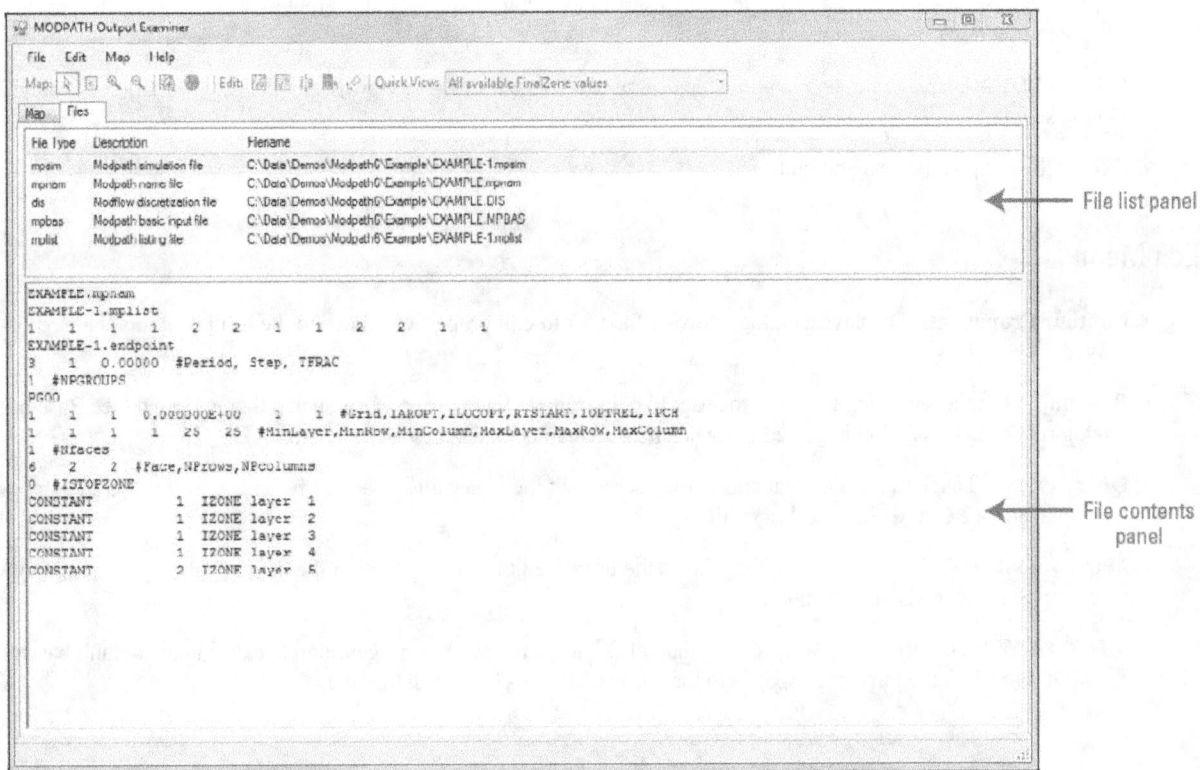

Figure 1–4.   Data files for example-1.mpsim viewed in files view.

Menus

The application is controlled by a number of commands and tools located in the File, Edit, and Map menus (figs. 1–5-1–7). The commands and tools contained in the Edit and Map menus also can be accessed directly on the toolbar.

File Menu

- **Open Simulation**—Displays a file-open dialog that allows the user to select a MODPATH simulation file.

- **Close Simulation**—Closes any open MODPATH simulation dataset and returns the application to its initial state with no dataset loaded.

- **New Basemap**—Displays a dialog to allow the user to create a new basemap (see Basemap Dialog section).

- **Load Basemap**—Displays a file-open dialog that allows the user to select a basemap to be loaded on the current map view.

- **Remove Basemap**—Removes a basemap from the current map view.

- **Save Basemap**—Saves the current state of a loaded basemap.

- **Save Basemap As**—Saves the current state of a loaded basemap as a new basemap name.

- **Export Shapefiles**—Displays a dialog to allow users to export grid and particle output as shapefiles (see Export Shapefile Dialog section).

- **Print to PDF**—Displays a dialog to allow the user to save the current map view in Adobe PDF format (see Print to PDF Dialog section).

- **Exit**—Closes open datasets and quit.

Edit Menu

- **Contour Properties**—Displays a dialog allowing the user to edit properties related to the head contours that are displayed.

- **Basemap**—Displays a dialog allowing the user to edit properties of the current basemap that is loaded (see Basemap Dialog section). This option is grayed-out when no basemap is loaded.

- **Query Filter**—Displays a dialog that allows the user to edit the query filter that will be used to select the data that will be displayed (see Query Filter Dialog section).

- **Map Symbology**—Displays a dialog that allows the user to edit the symbology of the data that is displayed in map view (see Map Symbology Dialog section).

- **MODFLOW Metadata**—Displays a dialog that allows users to specify grid georeference data and a default basemap for the MODFLOW dataset that corresponds to the current MODPATH simulation file.

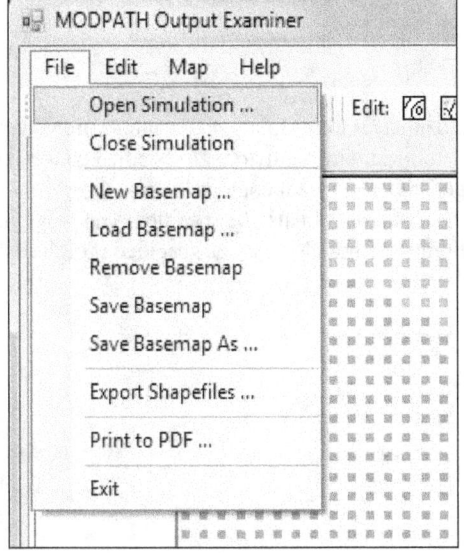

Figure 1–5. Screenshot of the File menu.

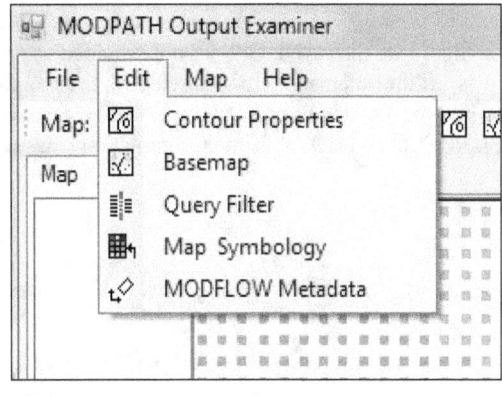

Figure 1–6. Screenshot of the Edit menu.

Map Menu

- **Pointer**—Sets the mouse cursor to be the default pointer tool. When this tool is active, the user can click on a head contour to display the contour level.

- **ReCenter**—Sets the mouse cursor to be the re-center tool. When this tool is active, the user can click on a location on the map to center the view on that location.

- **Zoom In**—Sets the mouse cursor to the Zoom-In tool. When this tool is active, the user can click on a map location to zoom-in by a factor of 1.5 and center the display on that location.

- **Zoom Out**—Sets the mouse cursor to the Zoom-Out tool. When this tool is active, the user can click on a map location to zoom-out by a factor of 1/1.5 and center the view on that location.

- **Zoom to Model Grid**—Resets the map display to the extent of the model grid.

- **Zoom to Full Extent**—Resets the map display to the full extent of the map data, including basemap layers.

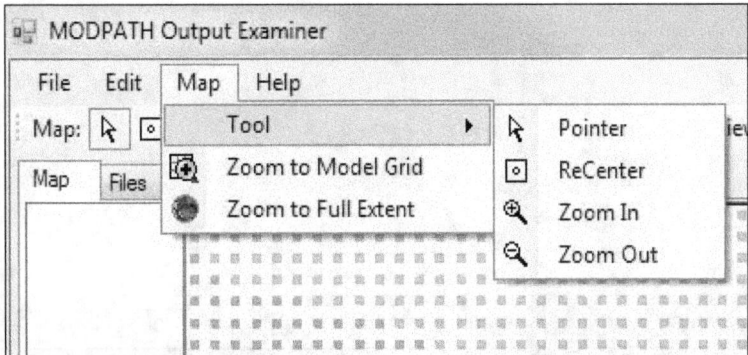

Figure 1–7. Screenshot of the Map menu.

Dialogs

Contour Properties Dialog

The Contour Properties dialog allows users to specify which model data layer to contour, what contours are generated, and how the contours are displayed. Options are provided to generate contours automatically or at a specified constant contour interval (fig. 1–8). If the automatic option is selected, the program will generate a constant contour interval that will produce 10 to 15 contour lines for the range of data in the model layer. Excluded values of head can also be specified. The program obtains the values of HNOFLO and HDRY from the simulation dataset, but additional values also may be specified if desired.

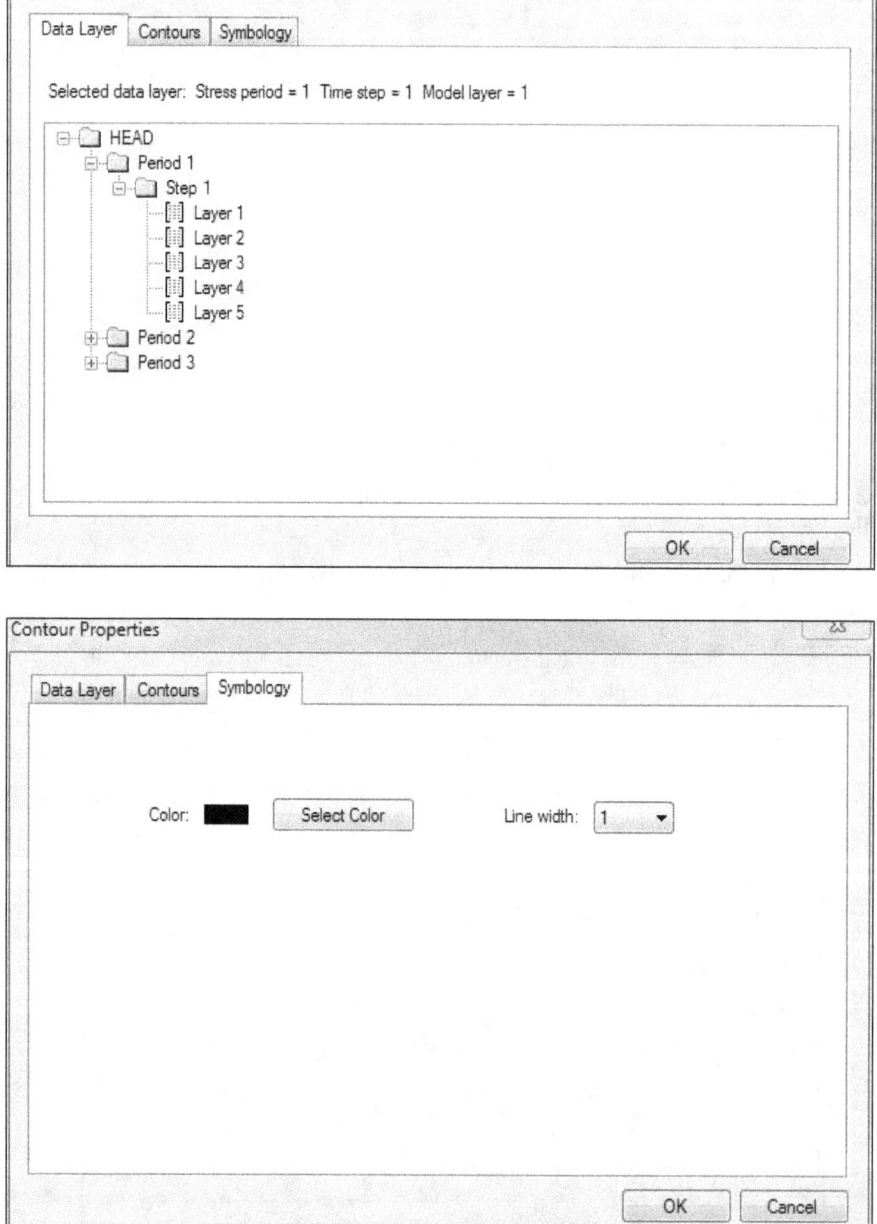

Figure 1–8   Screenshots of the Contour Properties dialog.

Basemap Dialog

MODPATH Output Examiner supports a simple basemap file structure that allows users to add reference data to the map view. Basemap layers correspond to shapefiles that must be created outside of the program. The basemap file stores the shapefile name, as well as the symbology and annotation data used to display the layer on the map and in the map explanation. Basemap files can be created from scratch by selecting the *New Basemap* option from the *File* menu (fig. 1–9). All shapefiles included in a basemap must be located in the same folder as the basemap file.

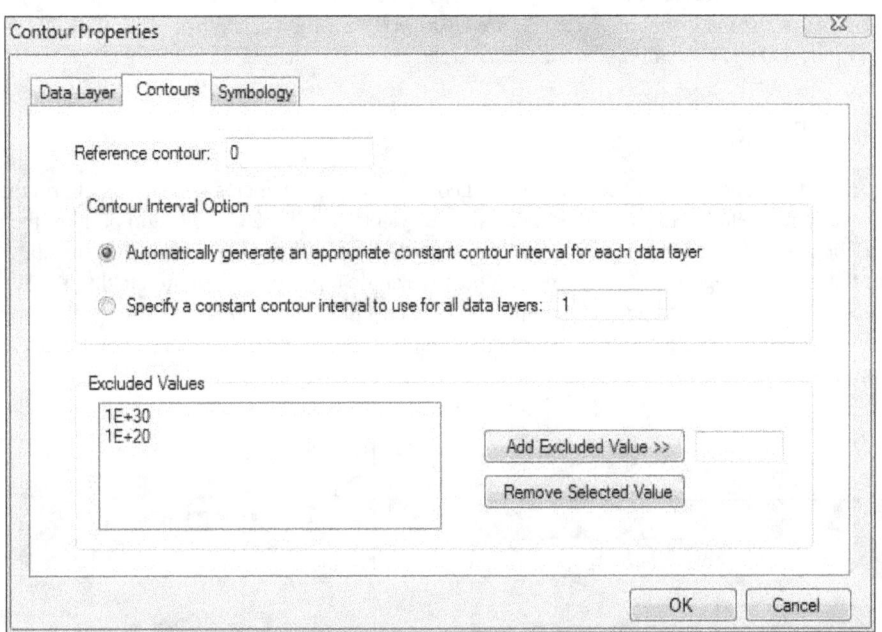

Figure 1–8   Screenshots of the Contour Properties dialog.—Continued

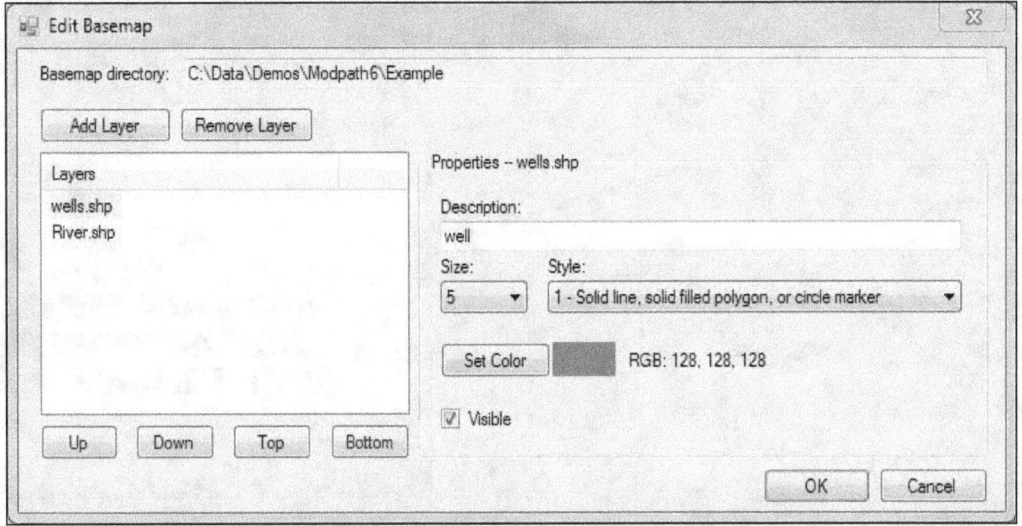

Figure 1–9.   Screenshot of the Basemap dialog.

Query Filter Dialog

A subset of the MODPATH output data can be displayed in map view by specifying a query filter that is applied to the output data. Only the particle data that meets the criteria set in the query filter is displayed on the map. The query filter options are customized for each of the basic MODPATH output types. The example shown in figure 1–10 is for endpoint simulation output.

Map Symbology Dialog

The symbology used to display output in map view can be customized by setting the properties in the Map Symbology dialog. The options provided in the map symbology dialog are customized for the basic types of MODPATH simulation output. The example shown in figure 1–11 is for endpoint simulation output.

MODFLOW Metadata Dialog

The MODFLOW metadata dialog allows users to specify georeferenced coordinates for the model grid that allows the particle-tracking output to be shown relative to a real-world coordinate system (fig. 1–12). The origin point of the grid is located in the lower left corner or the grid when the grid is viewed in an unrotated state with row 1, column 1 in the upper-left corner. The grid angle is restricted to be in the range -90 to 90 degrees. A positive angle represents a counterclockwise rotation. The angle may be specified explicitly, or by specifying a second point along the baseline of the grid and allowing the program to compute the angle. To use the option to compute the angle, click the button labeled Calculate Grid Angle and enter the coordinates of the second point in the pop-up dialog.

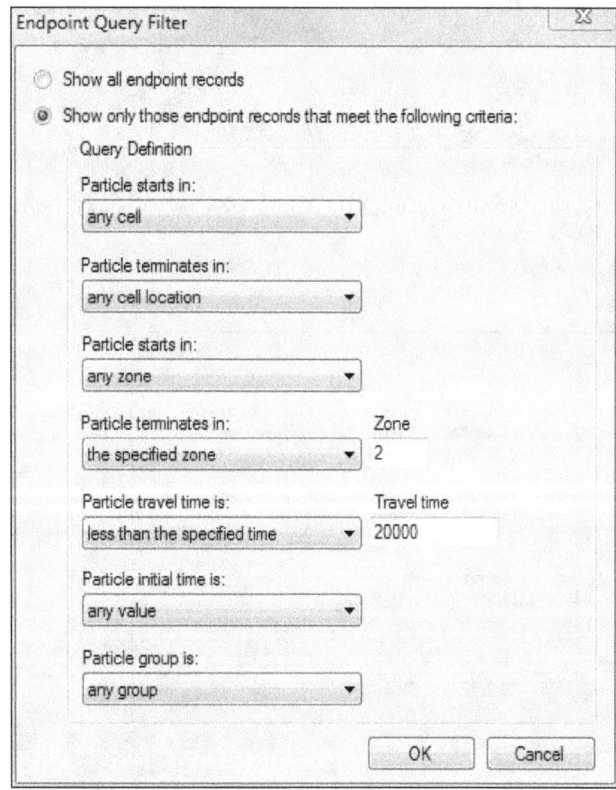

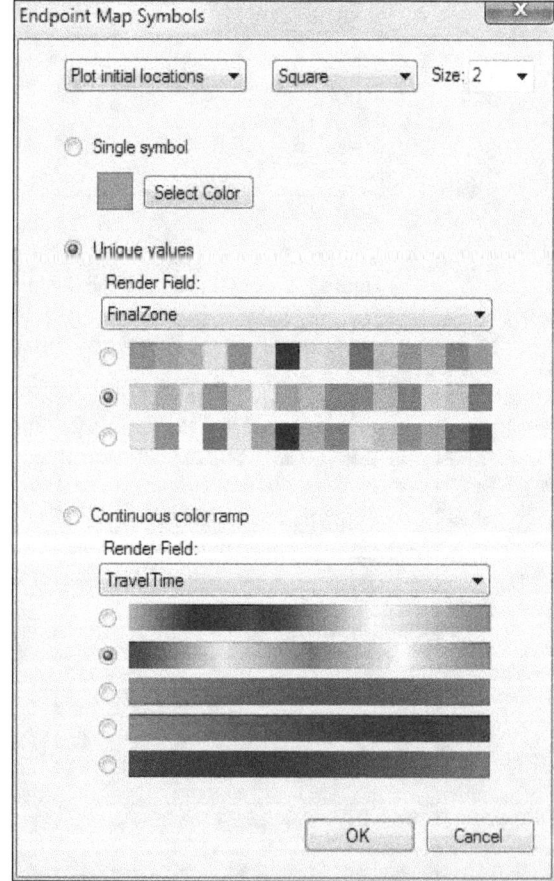

Figure 1–10.   Screenshot of the Endpoint Query Filter dialog.

Figure 1–11.   Screenshot of the Endpoint Map Symbology dialog.

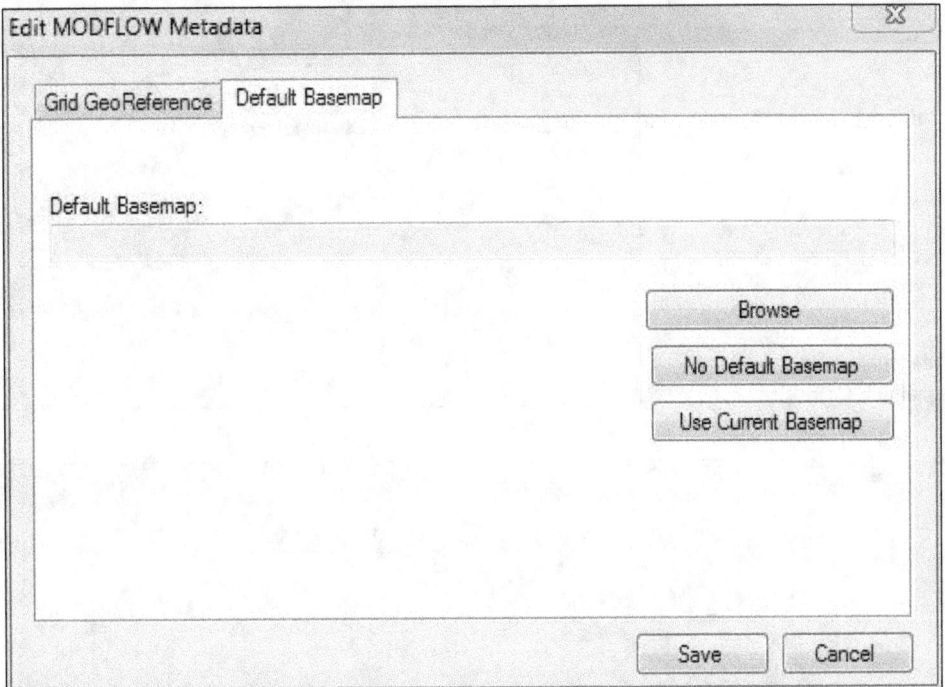

Figure A-12. Screenshots of the MODFLOW Metadata dialog.

A default basemap may also be specified as part of the MODFLOW metadata. If a default basemap is specified for a MOD-FLOW dataset, it will be loaded each time a MODPATH simulation is loaded that is based on that MODFLOW simulation.

The MODFLOW metadata is associated with a MODFLOW dataset and is stored in a file with a basename that is the same as the MODFLOW DIS package file with " metadata" added an extension. In the examples shown here, the metadata is stored in a file named Example-1.dis metadata.

Export Shapefiles Dialog

Model grid and particle output data (such as endpoints, pathlines, and timeseries points) can be exported as shapefiles. Spatial data in the form of shapefiles can be imported into many data visualization programs. MODPATH Output Examiner picks some default filenames that may be changed by the user (fig. 1–13). To maximize the portability of the shapefiles, the base names specified should not contain space or period characters. If a georeferenced grid has been defined and loaded for the simulation, the shapefiles will reflect the georeferenced coordinate system.

Print to PDF Dialog

The current map view can be saved to an Adobe PDF file (fig. 1–14). Optional title and description lines can be specified that will appear as header information in the PDF output file.

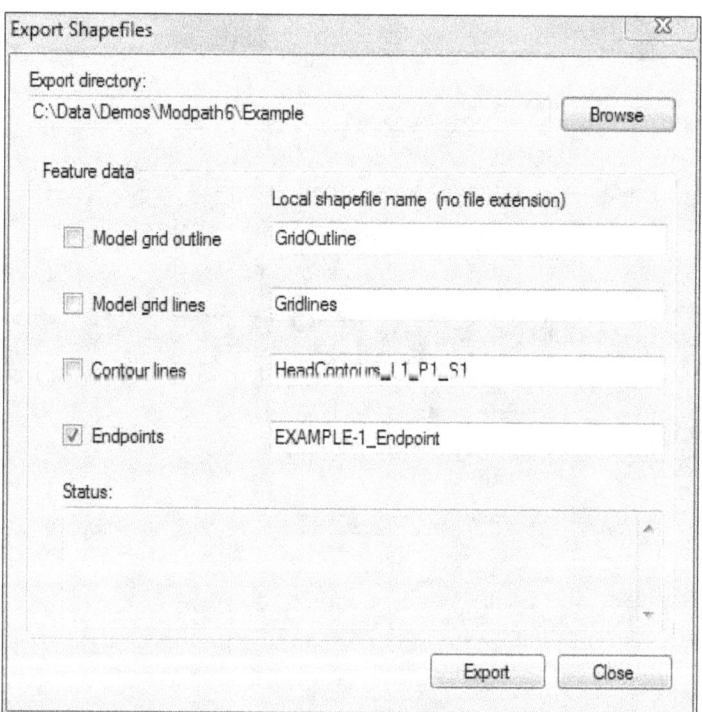

Figure 1–13. Screenshot of the Export Shapefiles dialog.

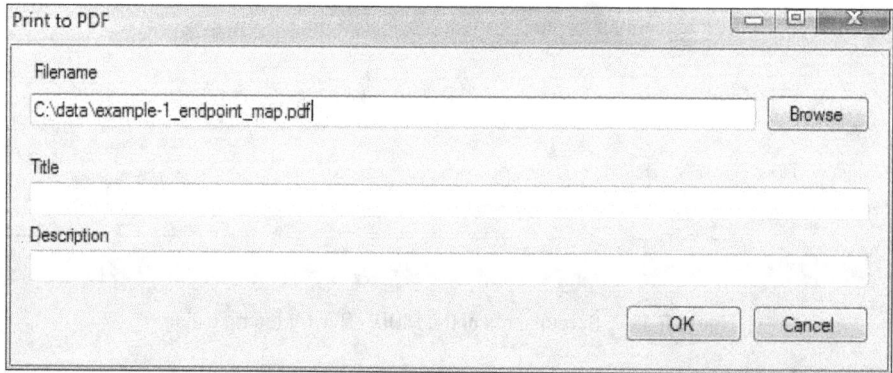

Figure 1–14. Screenshot of the Print to PDF dialog.

www.ingramcontent.com/pod-product-compliance
Lightning Source LLC
Chambersburg PA
CBHW081608170526
45166CB00009B/2877